MATLAB SIMULINK 使用入門 科技英文閱讀

盧並裕 著

博碩文化

Matlab Simulink使用入門：科技英文閱讀

作　　者／盧並裕
發 行 人／簡女娜
發行顧問／陳祥輝、竇丕勳
出　　版／博碩文化股份有限公司
網　　址／http://www.drmaster.com.tw/
地　　址／新北市汐止區新台五路一段112號10樓A棟
TEL / 02-2696-2869 • FAX / 02-2696-2867
郵撥帳號／17484299
律師顧問／劉陽明
出版日期／西元2012年7月初版一刷
建議零售價／350元
I S B N／978-986-201-615-2
博碩書號／EU31218

國家圖書館出版品預行編目資料

Matlab Simulink使用入門：科技英文閱讀 / 盧並裕作. -- 初版 -- 新北市：博碩文化, 2012.07
面；　公分
ISBN 978-986-201-615-2（平裝附光碟片）
1. Matlab（電腦程式）2. 英語 3. 讀本

312.49M384　　　101012216

Printed in Taiwan

序

大多數的科學軟體使用英文設定參數及使用說明，本書以 MatLab Simulink 這一套簡單的圖型軟體，說明如何以英文設定參數，最後並鼓勵讀者自行閱讀使用說明，進行程式設計，並練習科技英文閱讀。本書以大量的圖例說明協助英文閱讀，適合一般大學及技職院校之科技英文、專題製作、影像處理實習、及圖型軟體等課程之教科書及參考書及使用。本書之完成要感謝我在東南科技大學電子工程系的專題學生林洋良及許智博等同學協助驗證本書的程式。相信讀者在使用本書來操作軟體，會在科技與英文兩方面均能有所收穫。

盧並裕

於宜蘭三星

天主教聖母醫護管理專科學校

資訊管理科

2012/2/26

Preface

Scientific computer software usually presents their helps and instruction guides in English. Therefore, this book shows many examples of parameter settings in English for the graphical software of MatLab Simulink to aid the students' English reading. Furthermore, the translation of the sections in traditional Chinese was given by following the English ones to be convenient to the readers' comprehension. This book can be the textbook of the courses of image processing, graphical computer programming language, and English for engineering and technology. The examples have been verified by author's students who were Yang-Liang Lin, Zi-Po Hsu etc. in Department of Electronic Engineering, Tung-Nan University, Taipei, Taiwan. The author sincerely hopes to aid the readers to elevate the abilities in both technological level and English reading through the content of this book.

Benjamin Bing-Yuh Lu

in San-Sing, Yi-Lan

Department of Information Management,

Catholic St. Mary's Nursing, Management, and Medicine College,

Yi-Lan, Taiwan

Contents
目錄

Chapter 0　Introduction 導論

Chapter 1　Import and View the AVI Files 輸入與觀看 AVI 檔案

Chapter 2　Conversions 轉換

Chapter 3 Geometric Transformation 幾何轉換

Chapter 4 Morphological Operations 形態學上的操作

Chapter 5 Feature Extraction 特徵提取

Chapter 6 Image Analysis and Enhancement 圖像分析和強化

Chapter 7 Pixel Statistics 像素統計

Chapter 8 Pattern Matching 圖樣比對

Chapter 9 Image Compression 圖像壓縮

Chapter 10 Modulation Technology 調變的技術

Appendix A Solution 問題解答

Introduction
導論

0-1 Building the MatLab Simulink Model
如何建置 MatLab Simulink 的模型檔案

0-2 Creation of MatLab M-File
建置 MatLab M-File 程式

0-3 Where Can We Find the Default Images in the Examples of Video and Image Blockset?
我們能在哪裡找到本書中範例的影像檔?

0-1 Building the MatLab Simulink Model

Step 1 Double-click the “MATLAB” icon

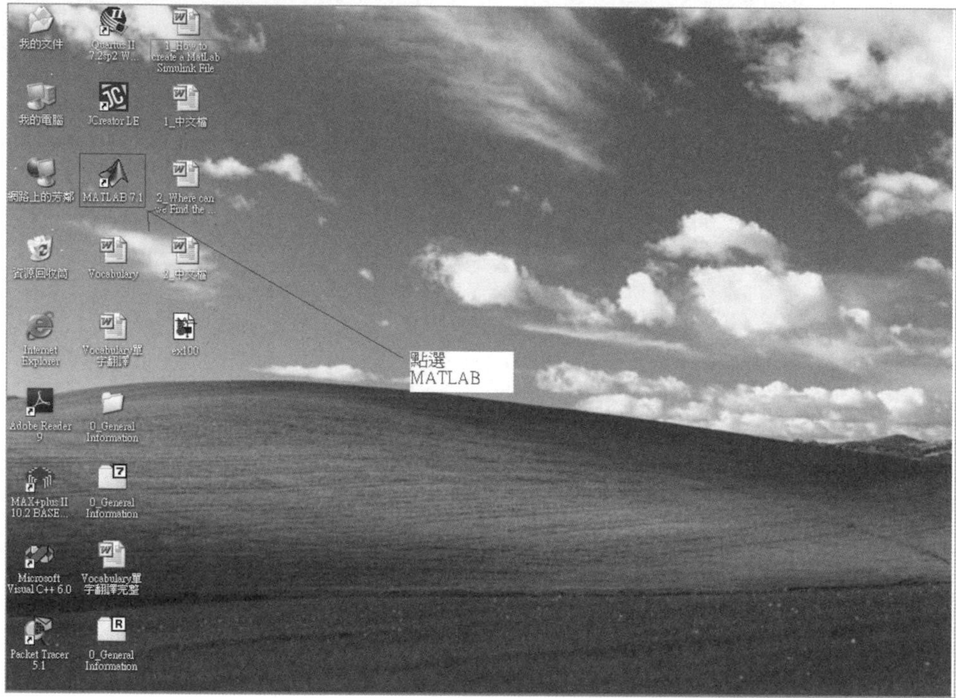

Step 2 Open MatLab

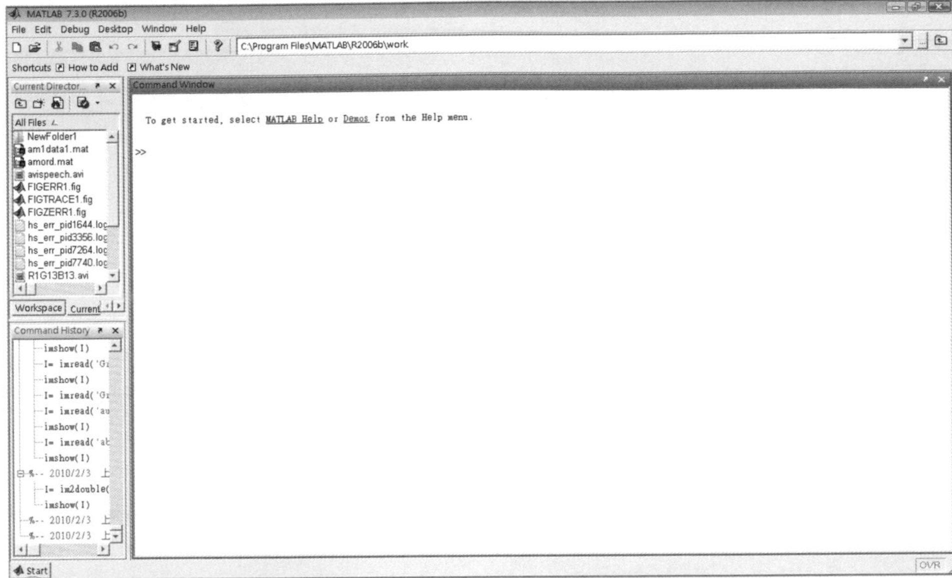

Step 3 Simulink icon

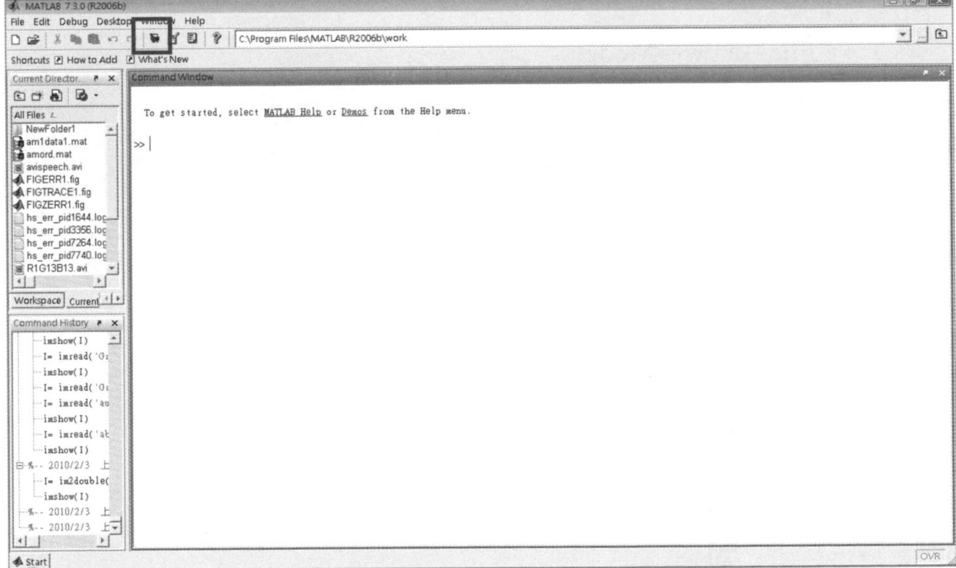

Step 4 Double click the icon, and the you will find the "Simulink Library Brower"

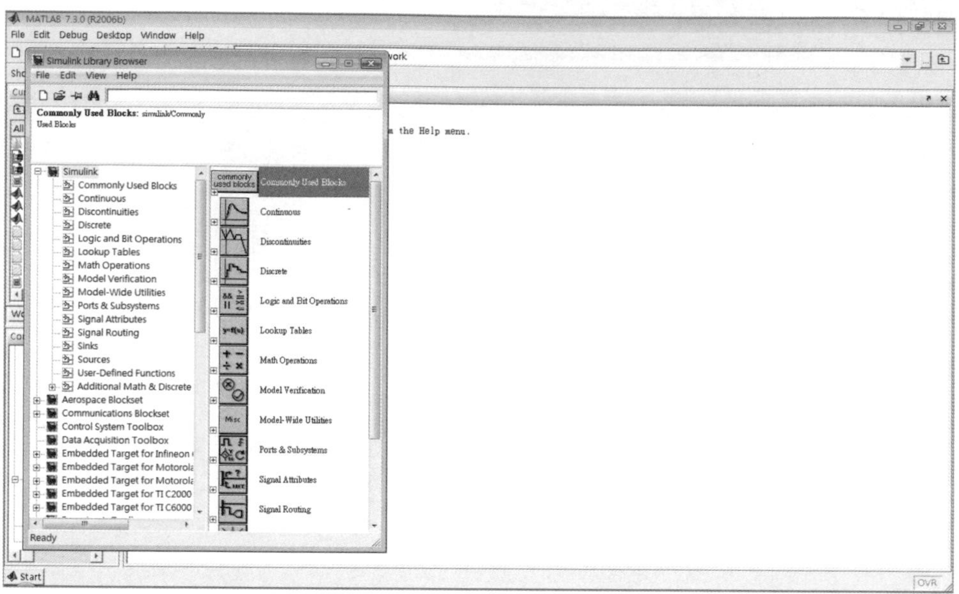

Step 5 Open file

File→New→Model

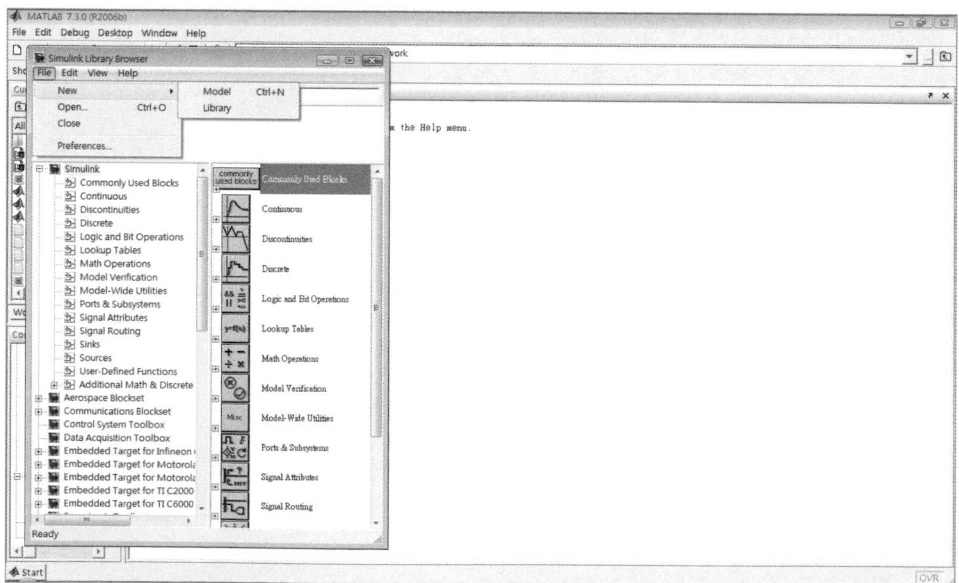

Step 6 Click the "Model", and you will find a new editor window to edit your model

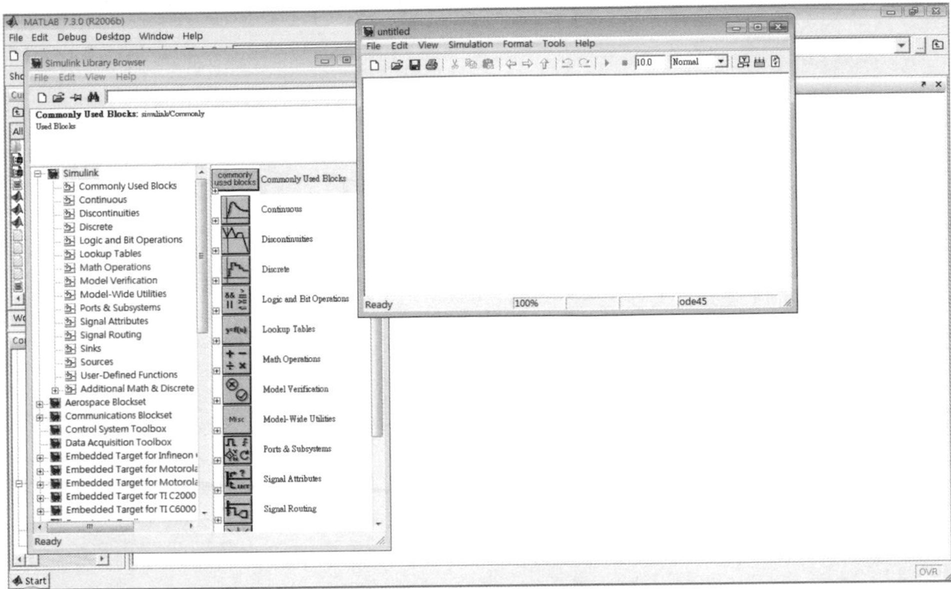

Step 7 Double click the "Source" button

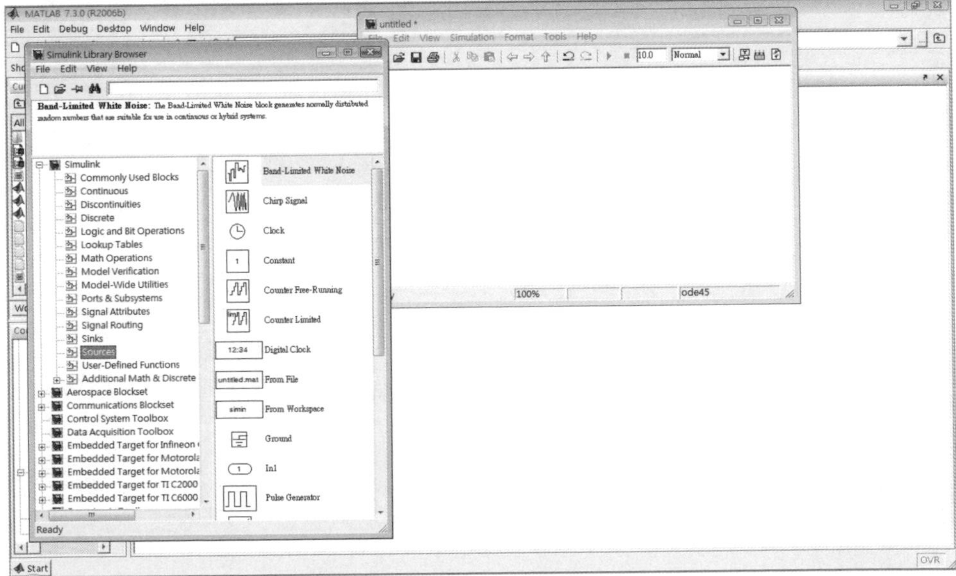

Step 8 Drag the icon of "Signal Generator" to the editor

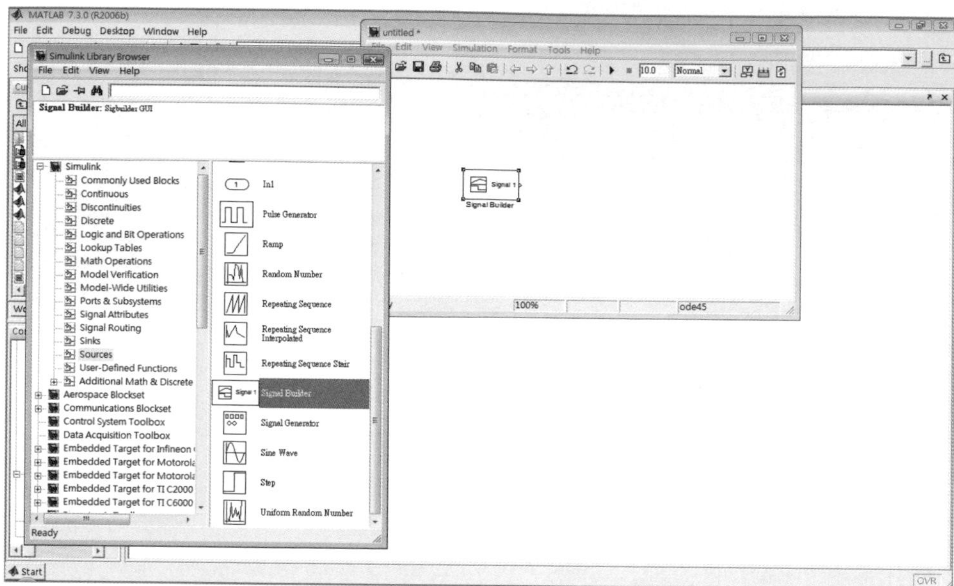

Step 9 Double click the "Sink" button

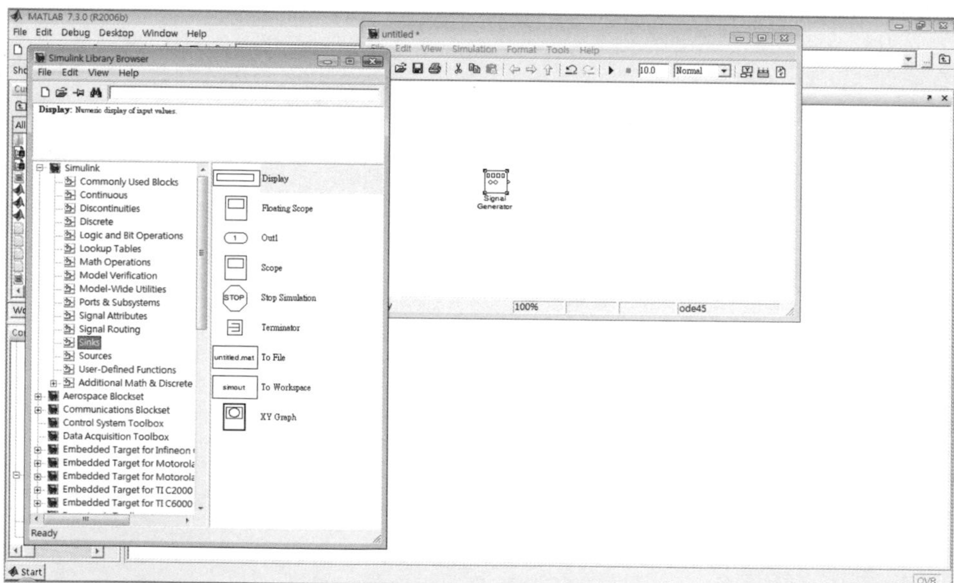

Step 10 Drag the icon of "Scope" to the editor

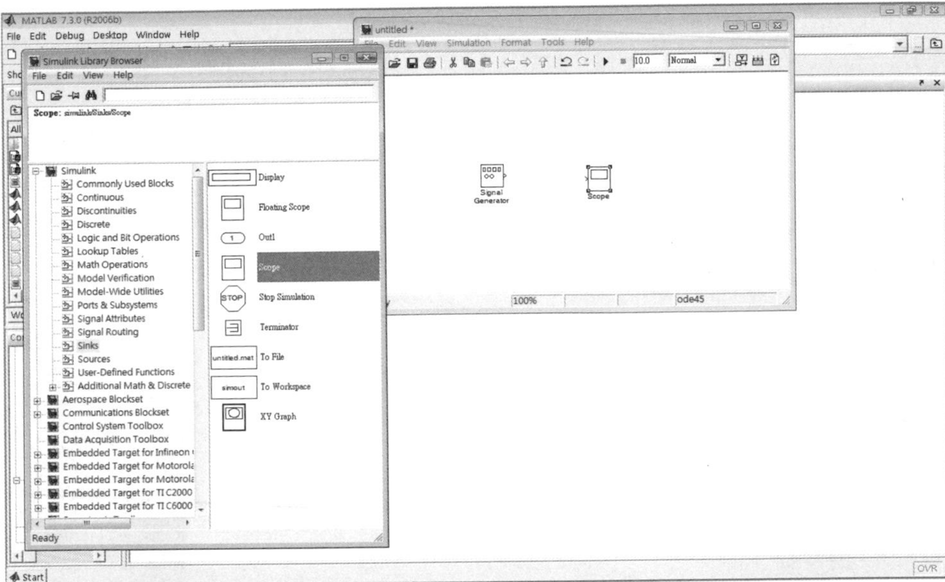

Step 11 Link the two blocks by keeping the press of left button on the mouse

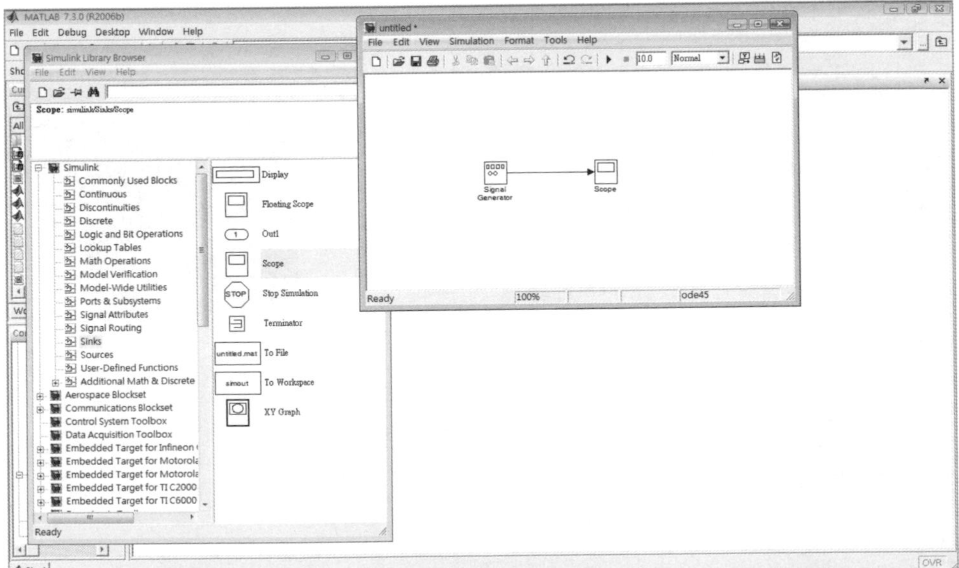

Step 12 Click the "play" button

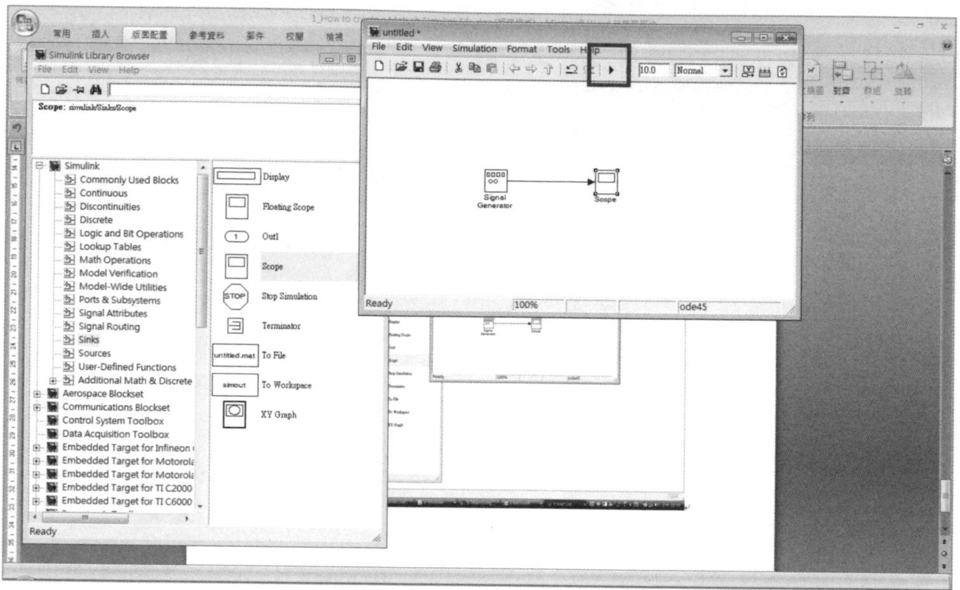

Step 13 Double click the "Scope" icon, and you may find the result

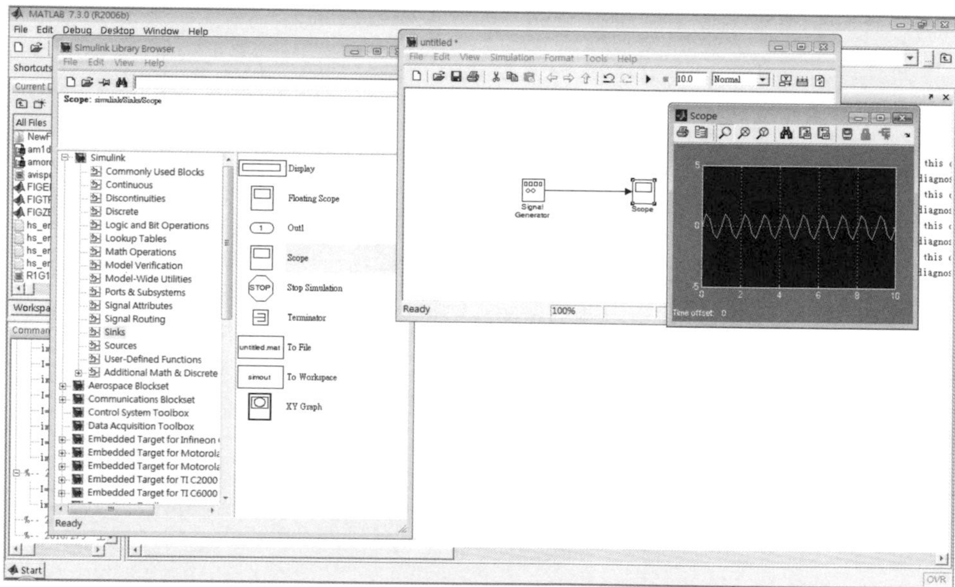

Step 14 Save the model

File→Save

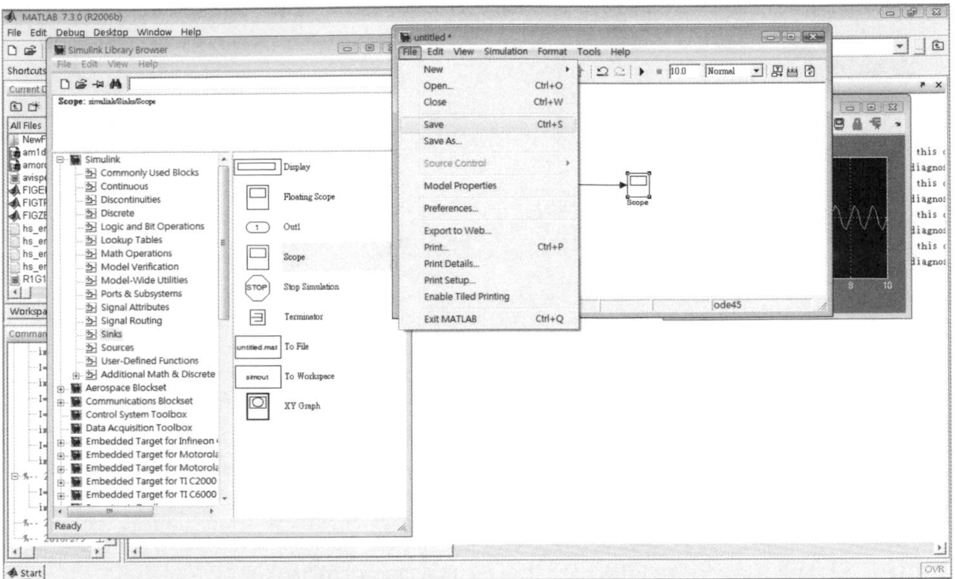

Step 15 Save the model as “ex100”

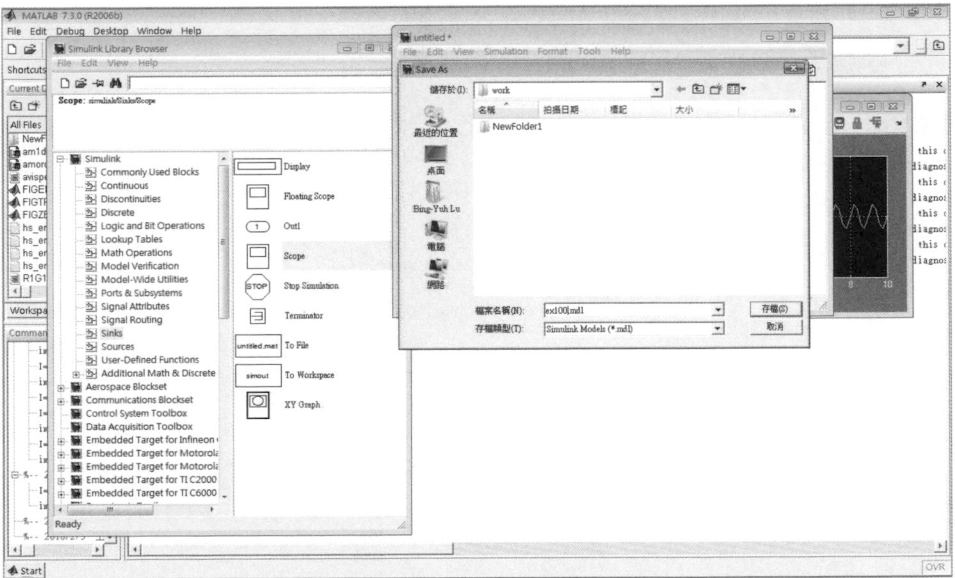

Step 16 Double click the “Signal Generator”

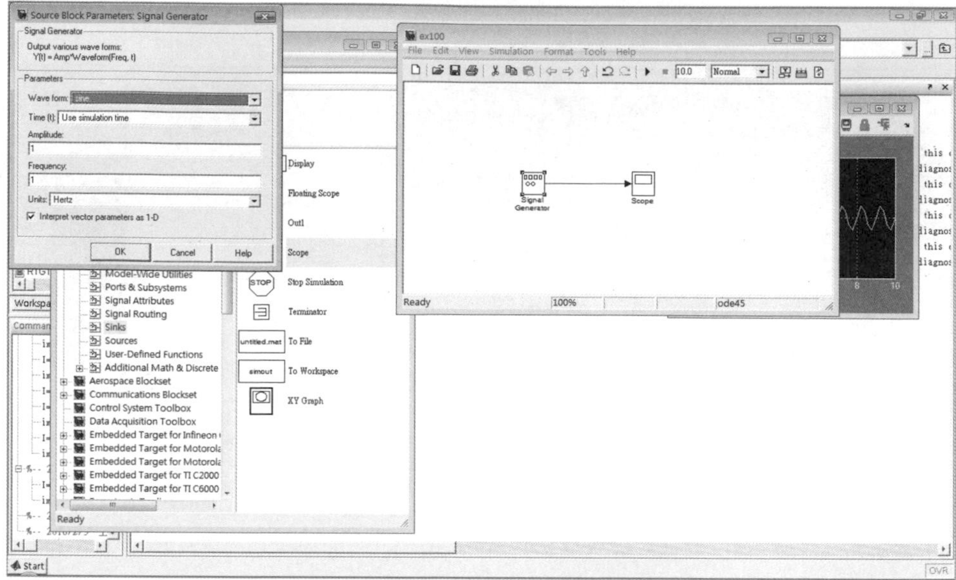

Step 17 Check the parameters of the signal generator:

Wave form=sine

Amplitude =1

Frequency=1

Step 18 Very good! You can make it! Go! Go! Go!

0-1 如何建置 MatLab Simulink 的模型檔案

Step 1 從桌面找到 MATLAB 的圖示並點選進入

Step 2 打開 MatLab

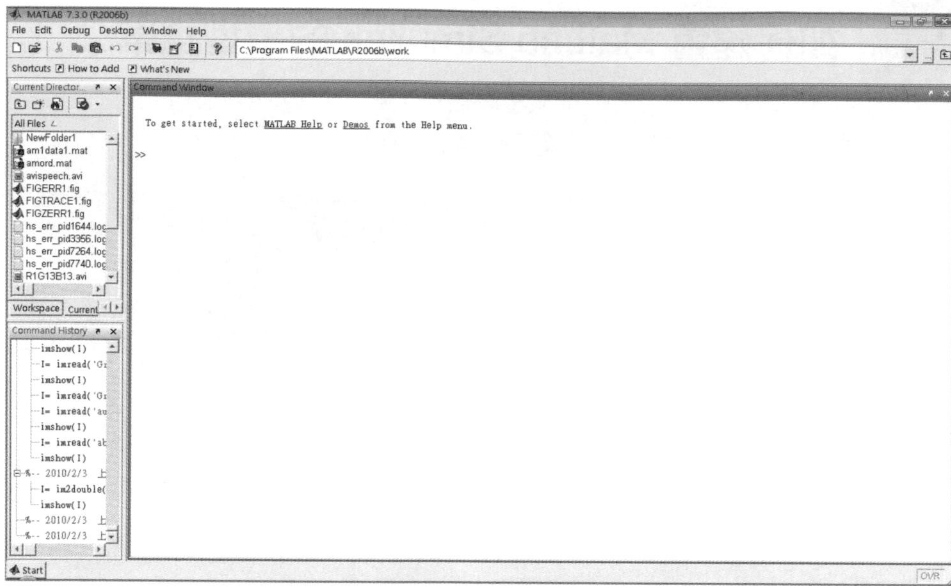

Step 3 Simulink 圖像

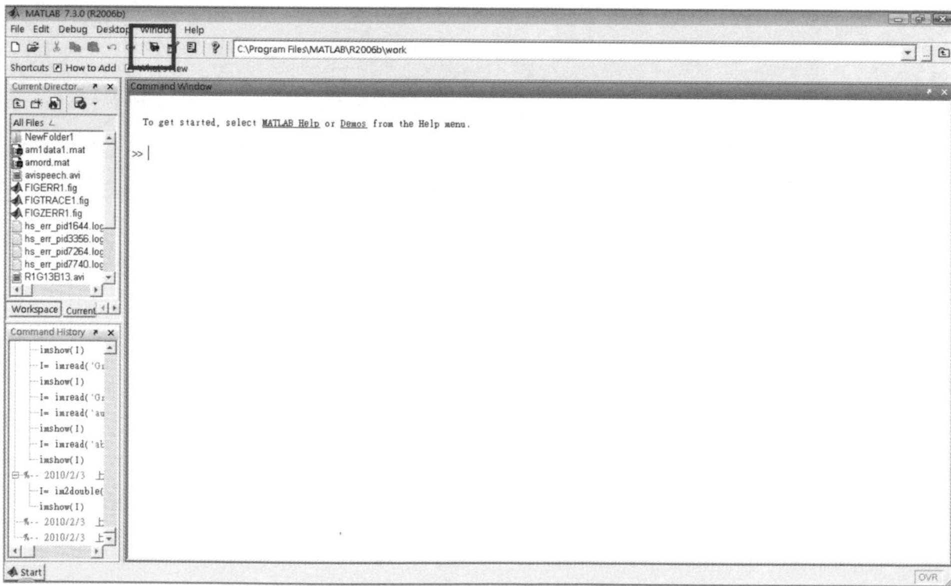

Step 4　雙點擊圖像，則出現“Simulink Library Brower (Simulink 元件庫總表)”

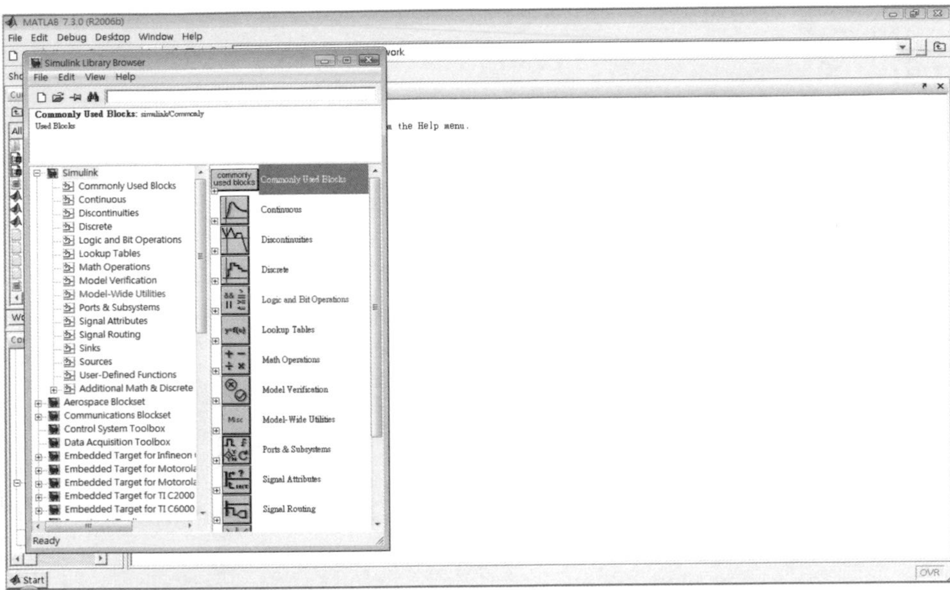

Step 5　打開檔案 File→New→Model(檔案→新的→模型)

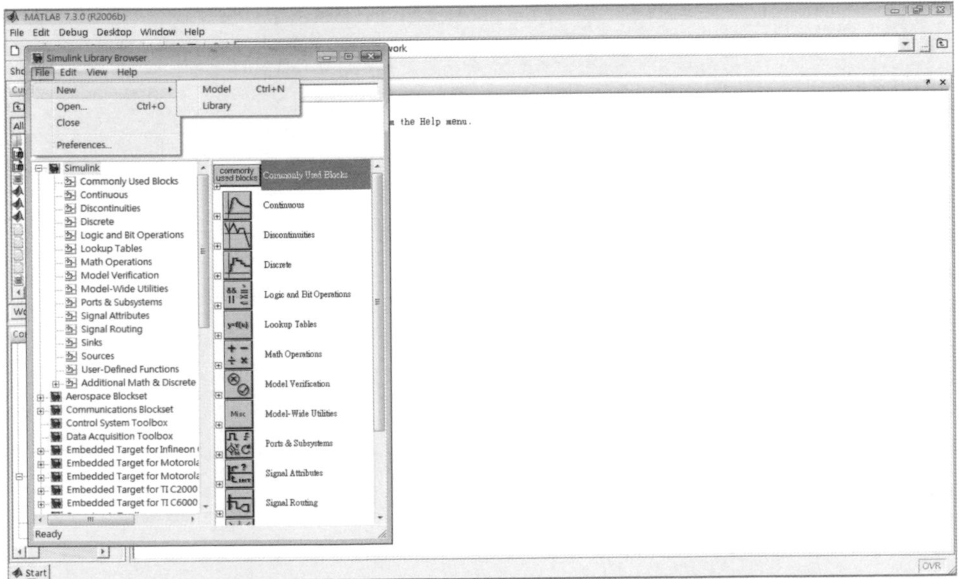

Step 6 按 " Model "，將發現一個新視窗來編輯你的模型

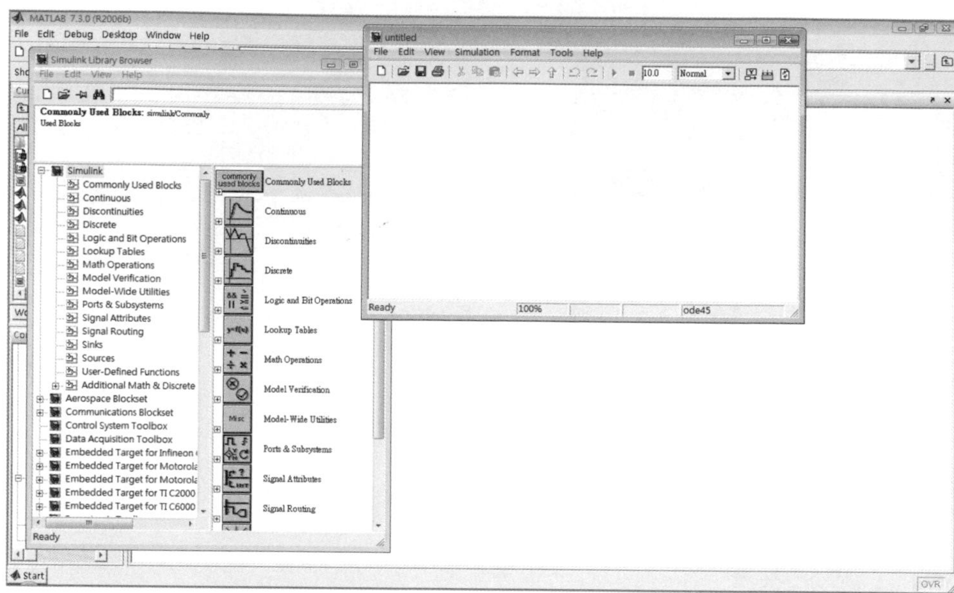

Step 7 雙點擊 " Source " 按鈕

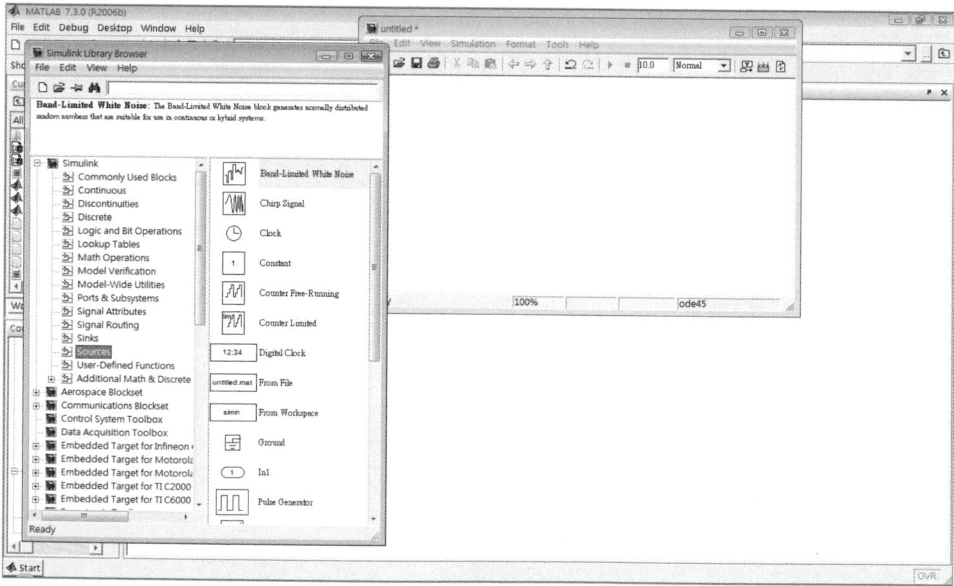

Step 8　拖曳 " Signal Generator " 的圖像到編輯視窗

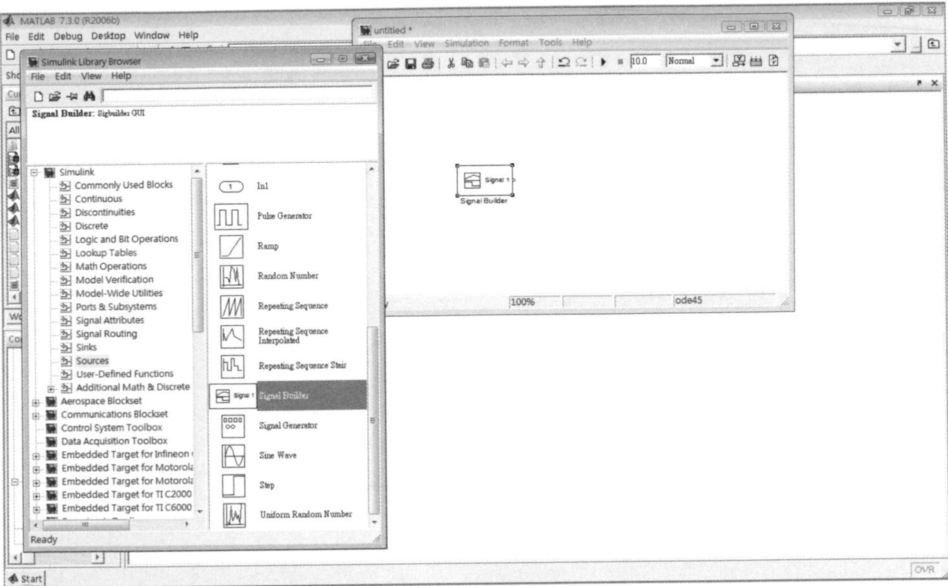

Step 9　點兩下 " Sink " 按鈕

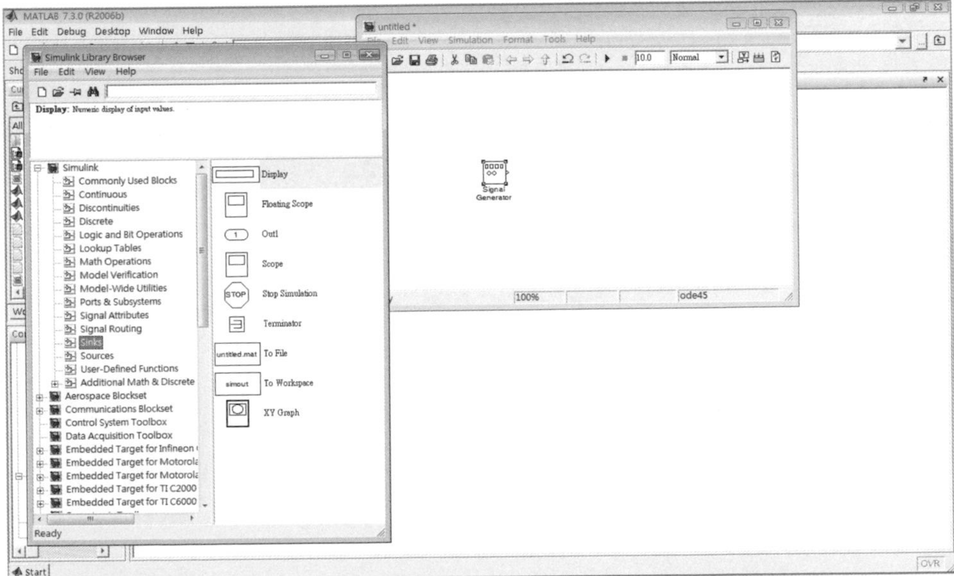

Step 10　拖曳 " Scope " 的圖像到編輯視窗

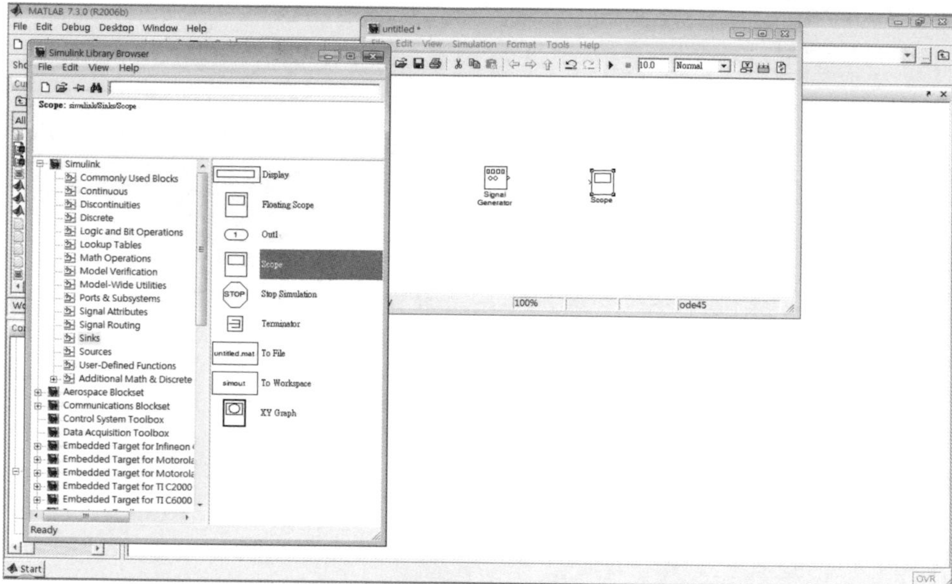

Step 11　按住滑鼠左鍵並拖曳來連接這二個元件

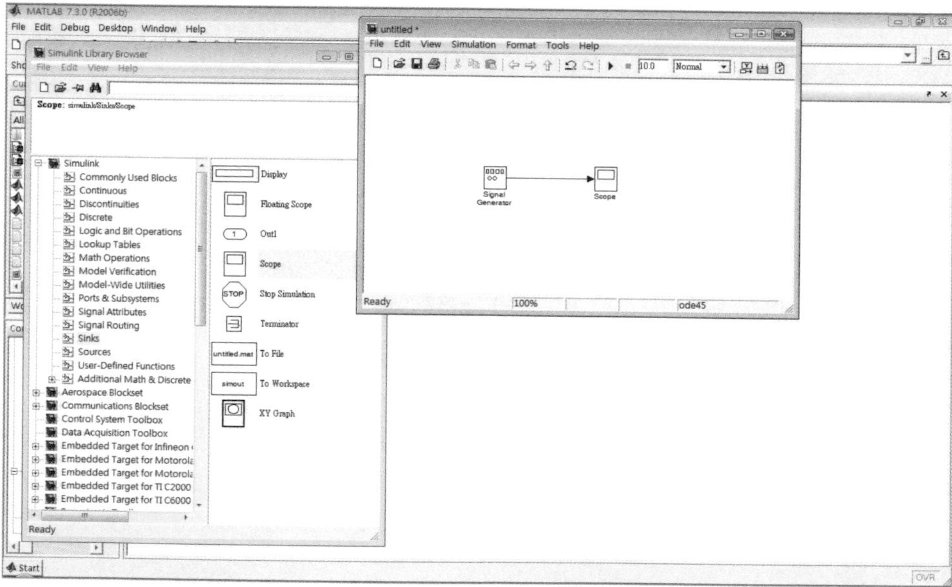

Step 12 按 " play " 按鈕

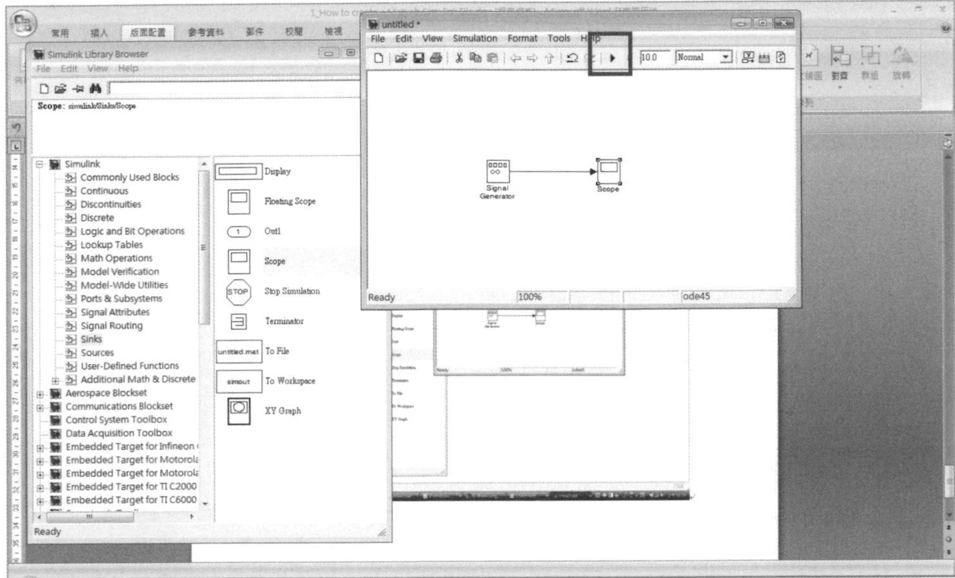

Step 13 按兩下 " Scope " 的圖像，你會發現

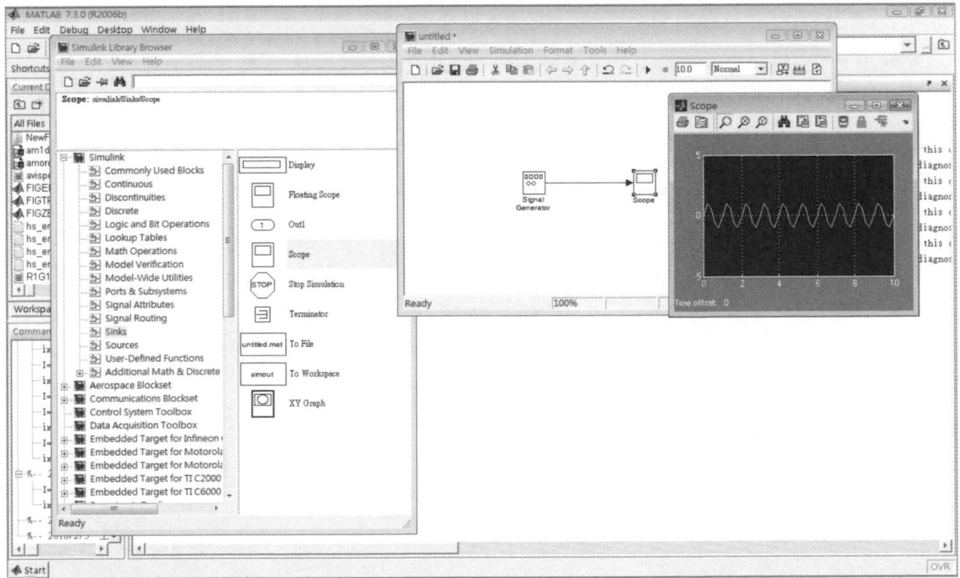

Step 14 儲存模型 File→Save(檔案→儲存)

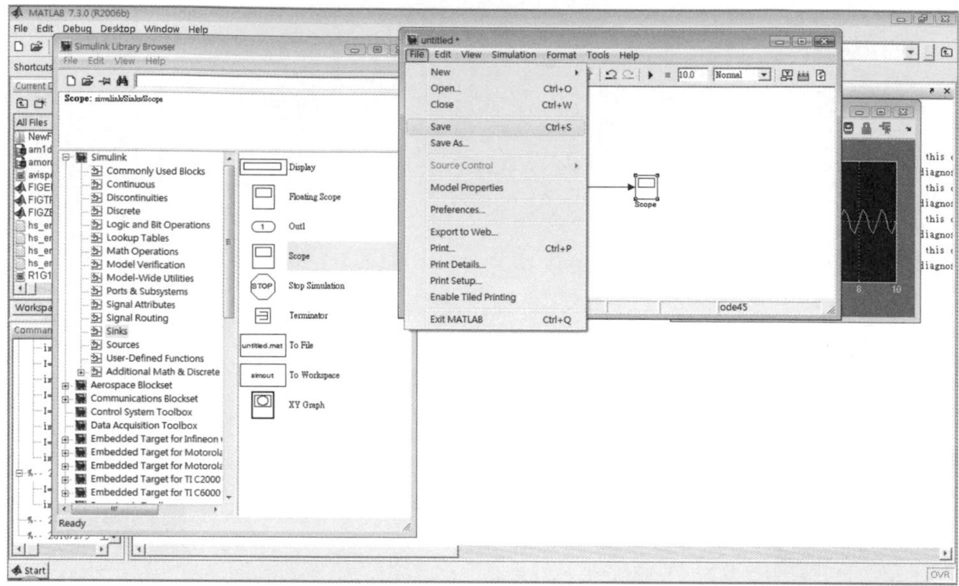

Step 15 儲存模型的名稱為＂ex100＂

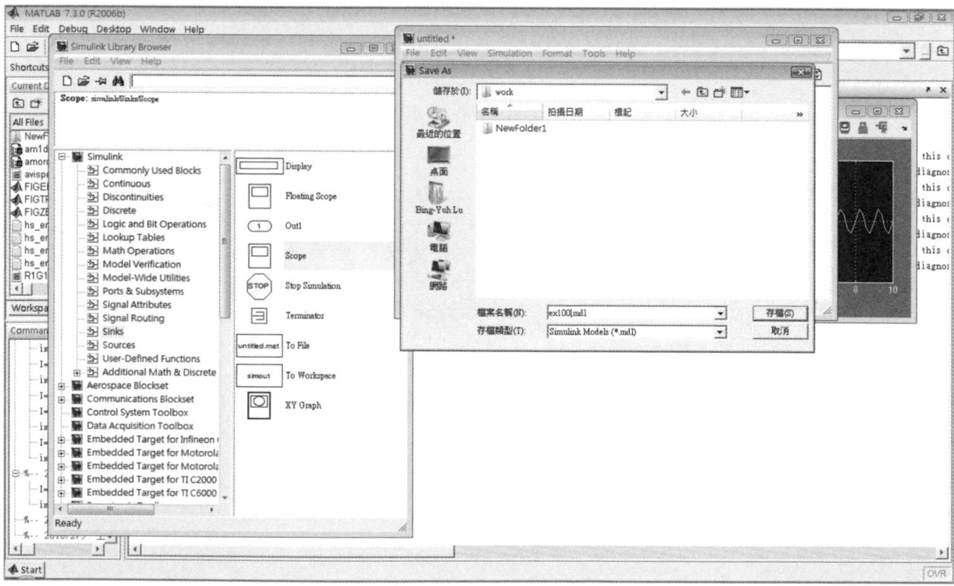

Step 16 按兩下 " Signal Generator "

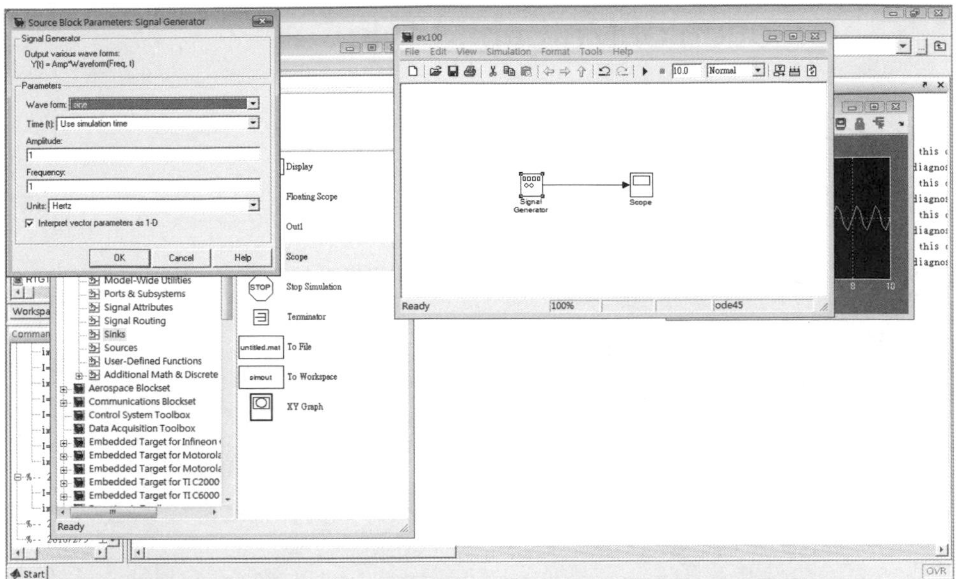

Step 17 檢查 signal generator 的參數

Wave form(波型)=sine(正弦波)

Amplitude(振幅) =1

Frequency(頻率)=1

Step 18 很好！你可以做到的！加油！

0-2 Creation of MatLab M-File

1. MatLab Interpreter

Step 1 Double-click the "MatLab" icon.

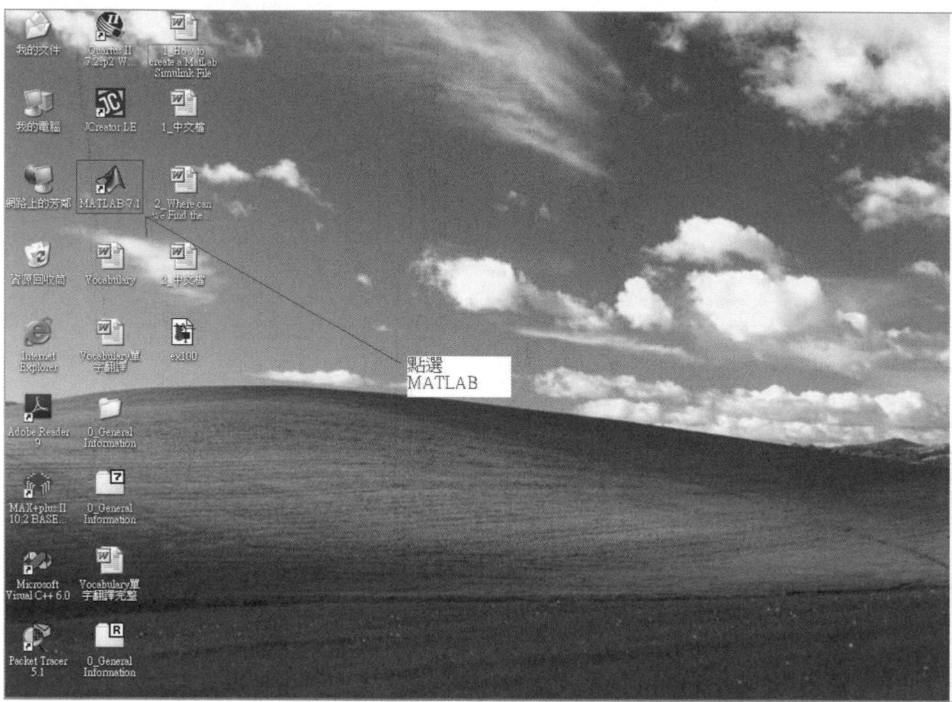

Step 2 Open MatLab

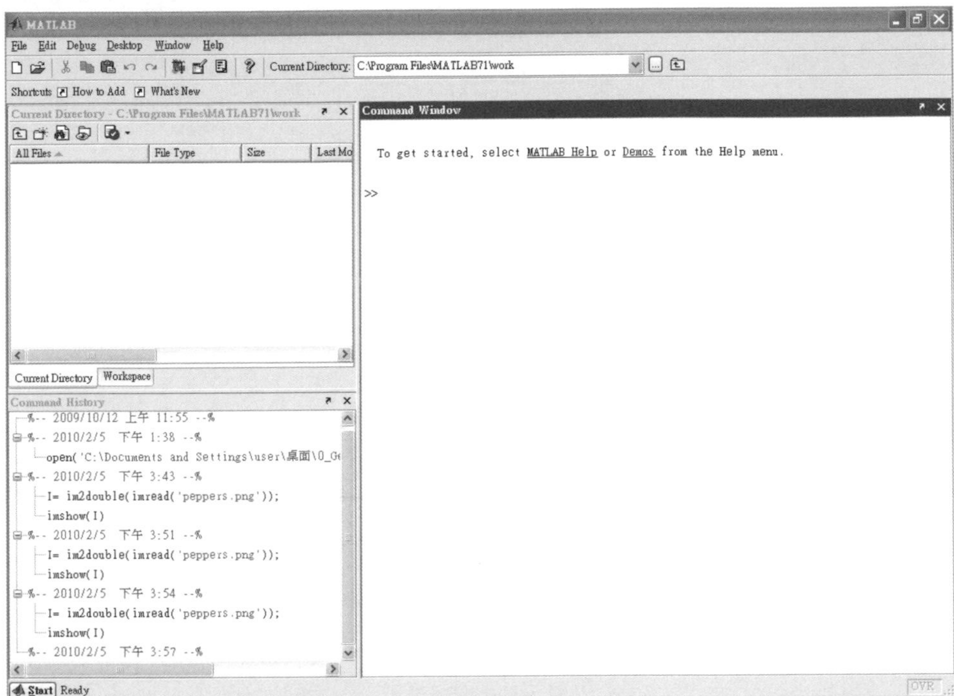

Step 3 Key in the following 2 instructions

```
I= im2double(imread('peppers.png'));

imshow(I)

%    "%" notes the comments of the program

% im2double : Transfer the format of image into "double"

% imread : Read an image file

% imshow : Show an image file
```

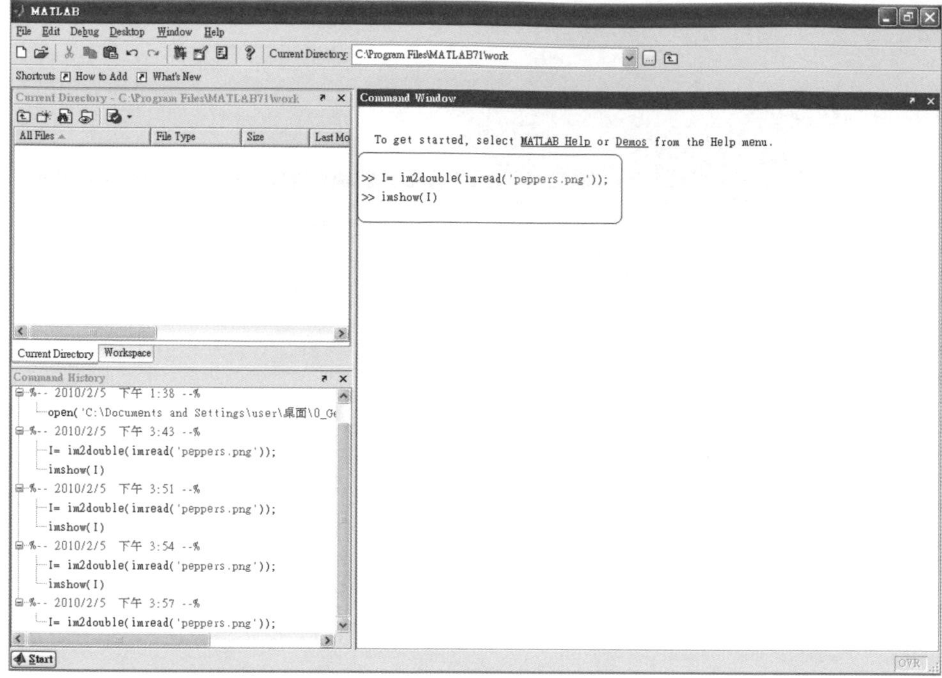

Step 4 Show the image

2. MatLab M-File

Step 1 Double-click the "MATLAB" icon.

Step 2 Open a file

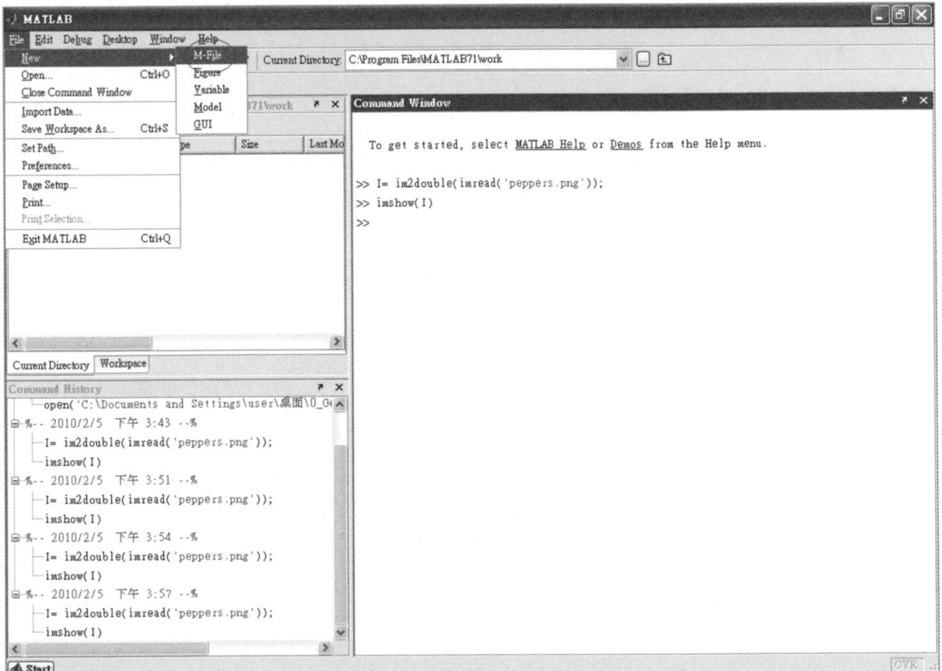

Step 3 Press" M-file", and you will find a new Windows for the programming editor

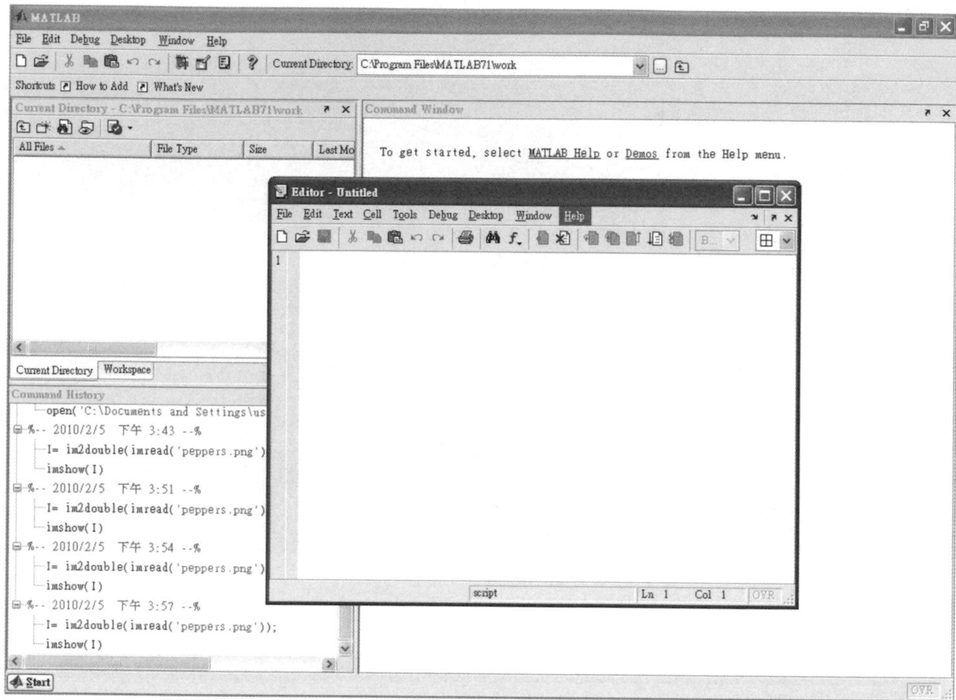

Step 4 Code the M-File

```
I= im2double(imread('peppers.png'));
imshow(I)
```

Editor - C:\Program Files\MATLAB71\work\pro100.m

File Edit Text Cell Tools Debug Desktop Window Help

```
I= im2double(imread('peppers.png'));
imshow(I)
```

Step 5 Save

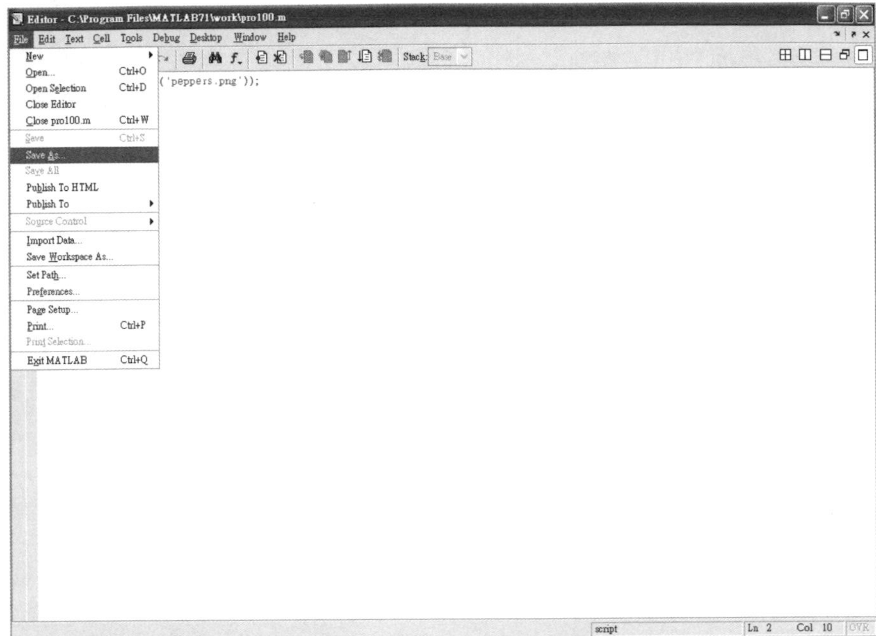

Step 6 The file name is " pro 100"

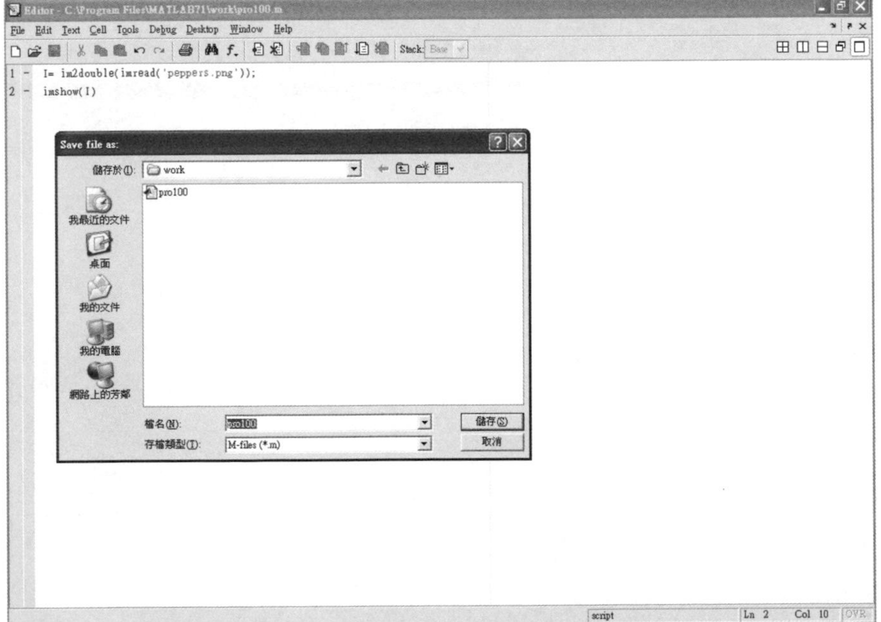

Step 7 Return the Command Windows

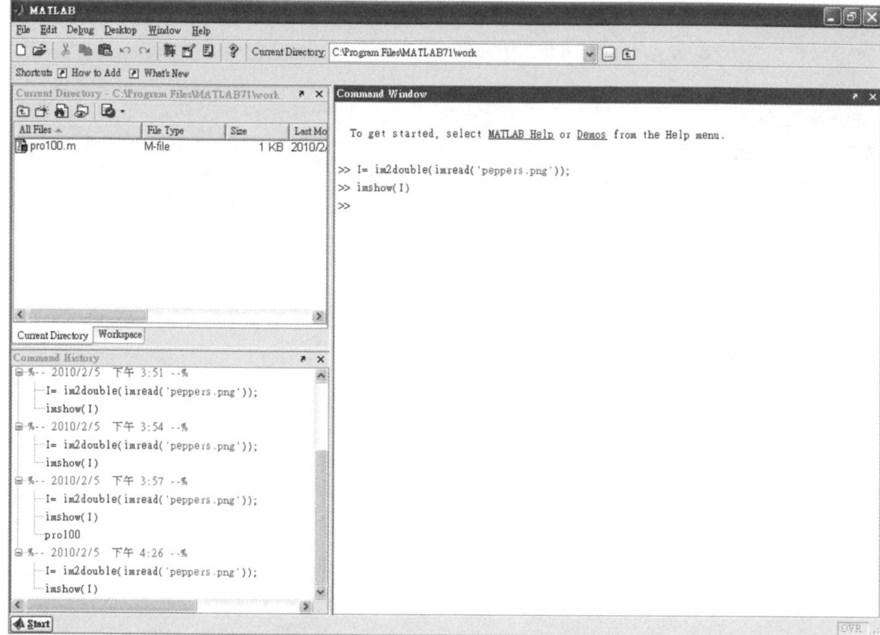

Step 8 Key in "pro100", and press Enter

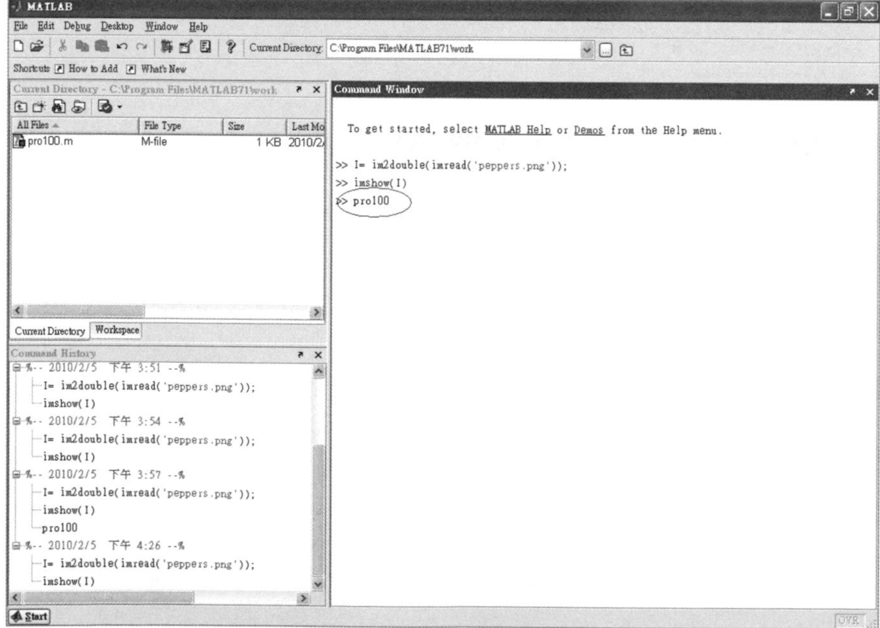

Step 9 You may find the same result as using the interpreter

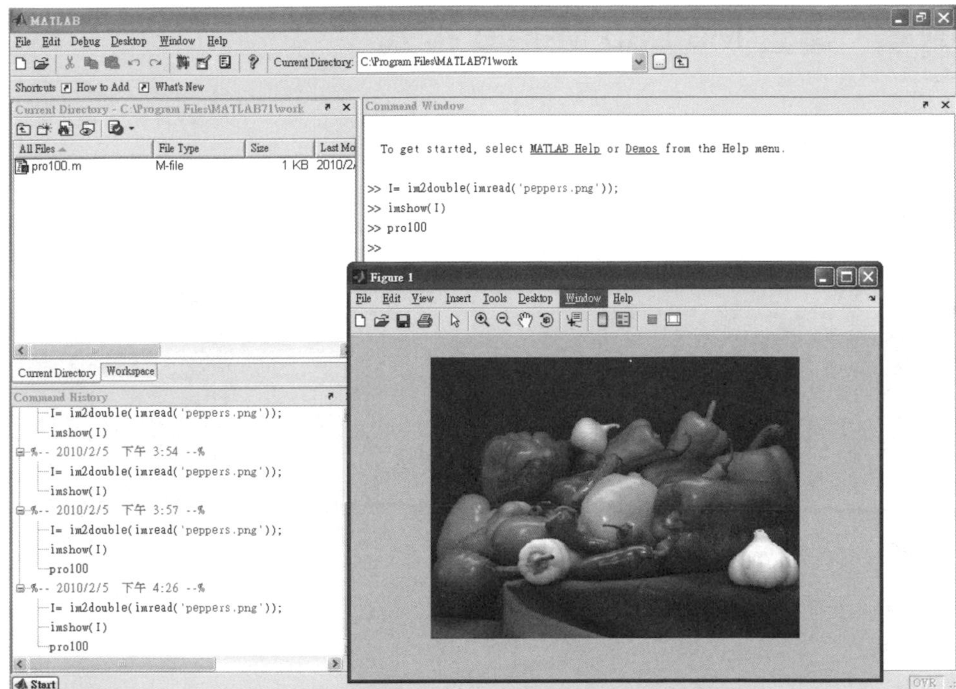

Step 10 Go! Go! You may make it!

0-2 建置 MatLab M-File 程式

1. MatLab 直譯器

Step 1 從桌面找到 MatLab 的圖示，並點選進入

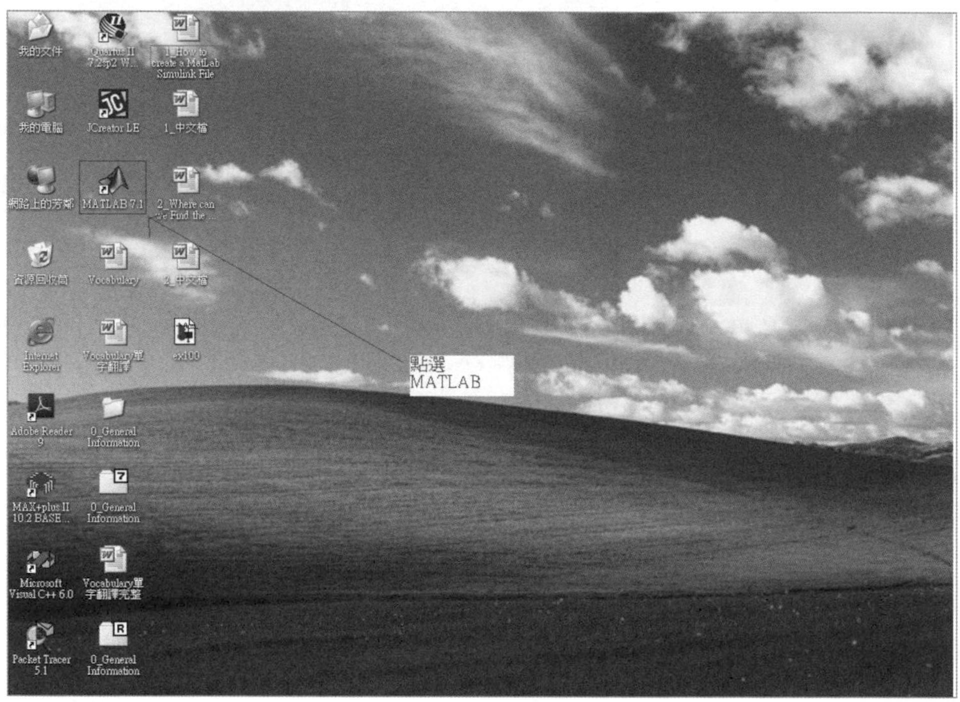

Step 2　打開 MatLab

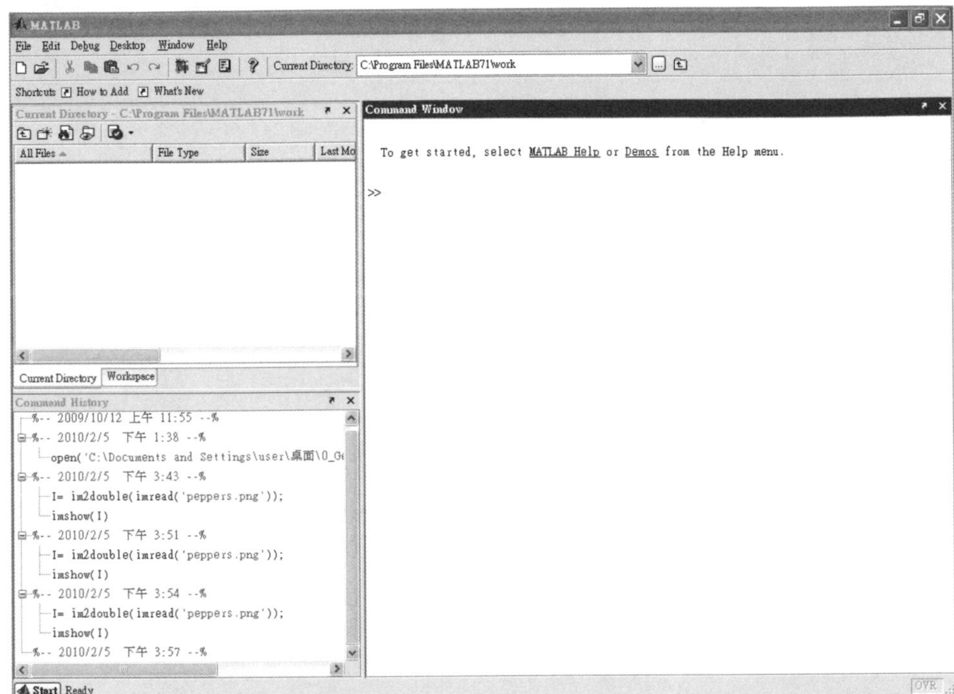

Step 3　在視窗上程式輸入

```
I= im2double(imread('peppers.png'));
imshow(I)
%  “%” 為註解的符號
% im2double：把圖檔轉成兩倍的倍精數
% imread：讀一個圖檔
% imshow：把圖顯示出來
```

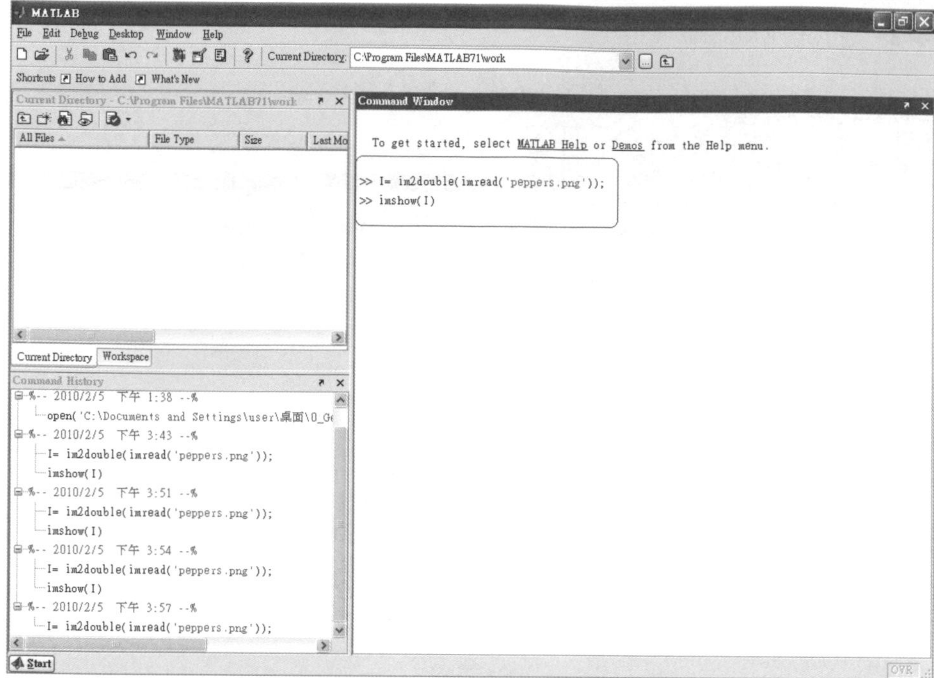

Step 4 按 Enter 把圖展示

2. 使用 MatLab 編輯器寫程式

Step 1 從桌面找到 MatLab 的圖示，並點選進入

Step 2 打開檔案－File→New→M-file(檔案→新的→M 檔案)

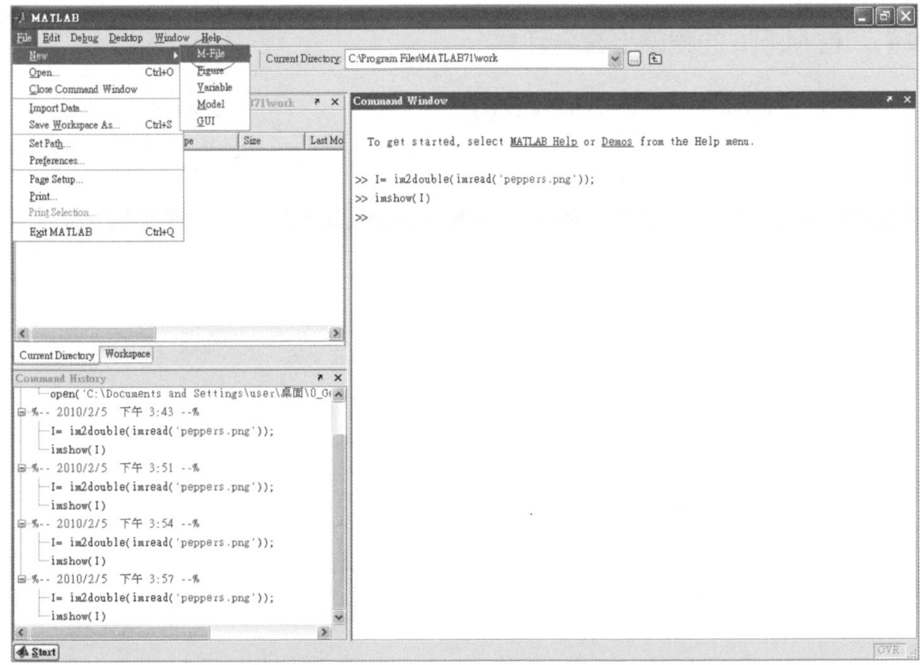

Step 3 按 " M-file "，你將發現一個新的視窗可以編輯你的程式

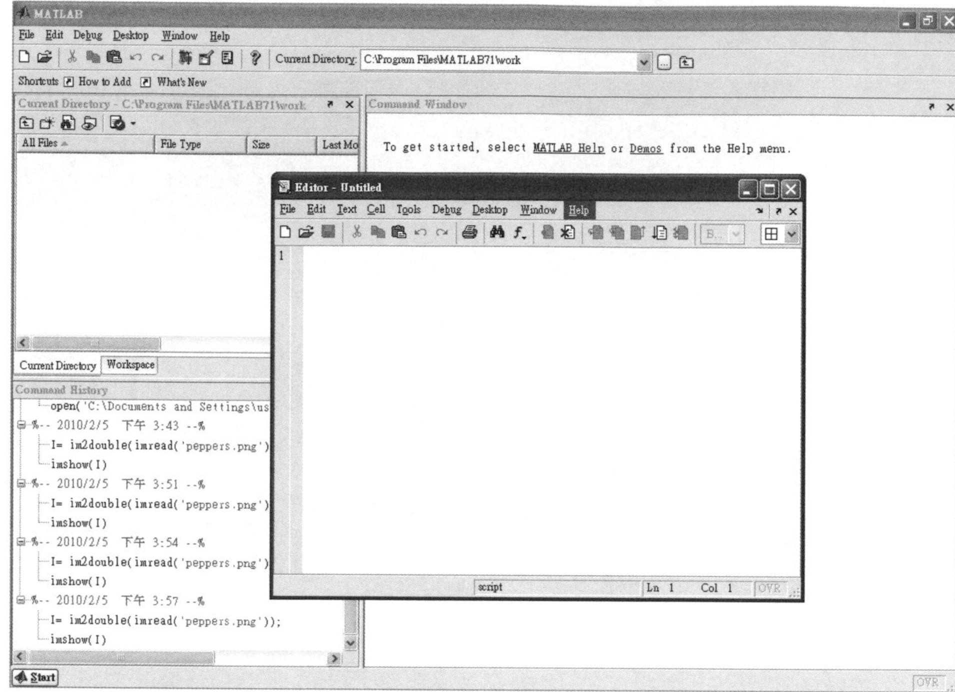

Step 4 在 M-File 的視窗上寫入程式

```
I= im2double(imread('peppers.png'));

imshow(I)
```

Editor - C:\Program Files\MATLAB71\work\pro100.m

File Edit Text Cell Tools Debug Desktop Window Help

Stack: Base

```
1 - I= im2double(imread('peppers.png'));
2 - imshow(I)
```

Step 5 儲存程式－File→Save as

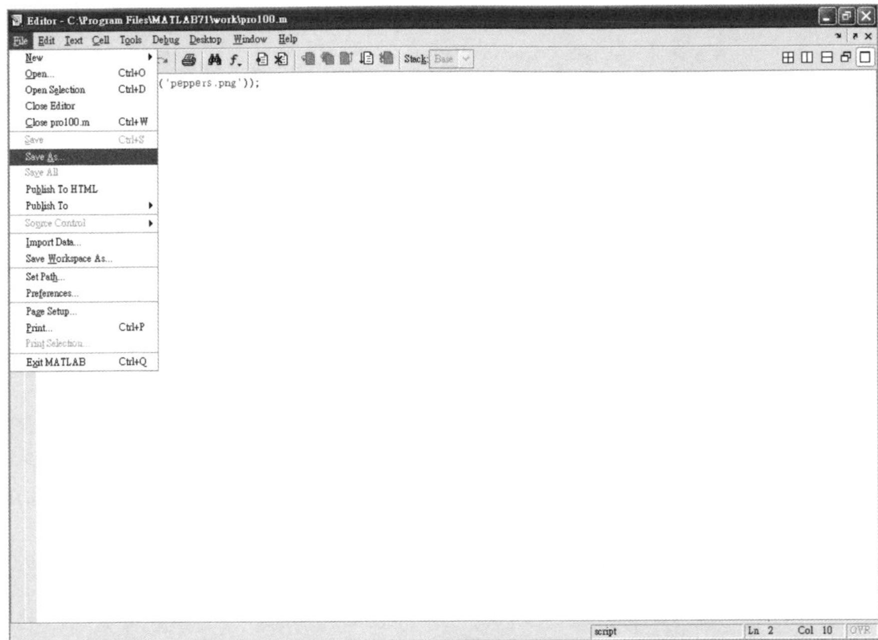

Step 6 儲存程式的名稱為 " pro100 "

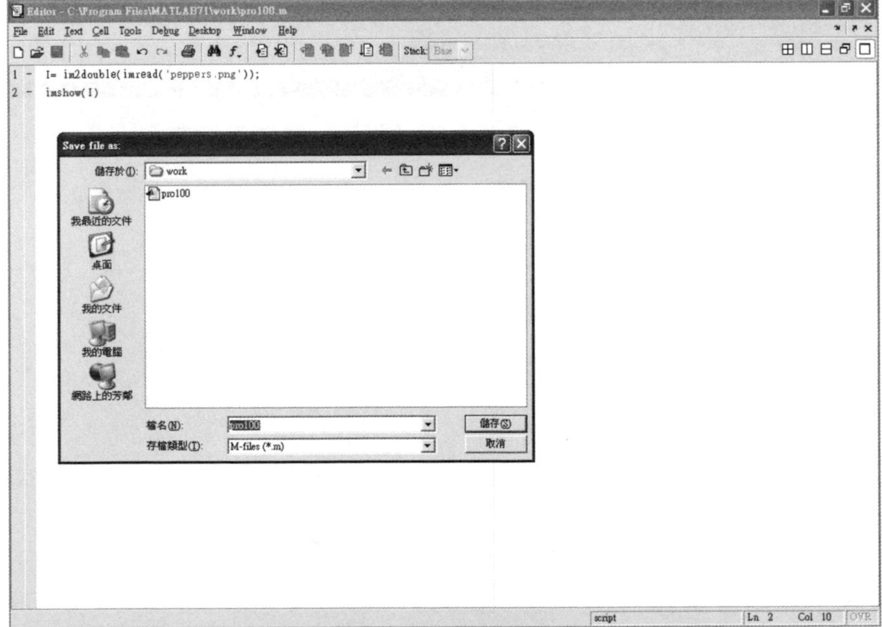

Step 7 回到 MatLab 剛開始打開的視窗(指令視窗)

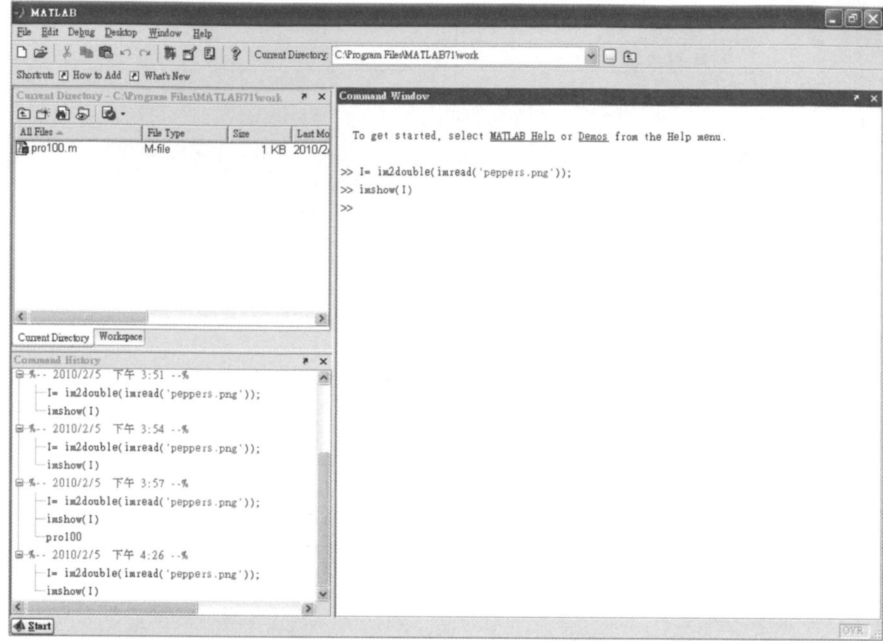

Step 8 輸入"pro100"並按 Enter

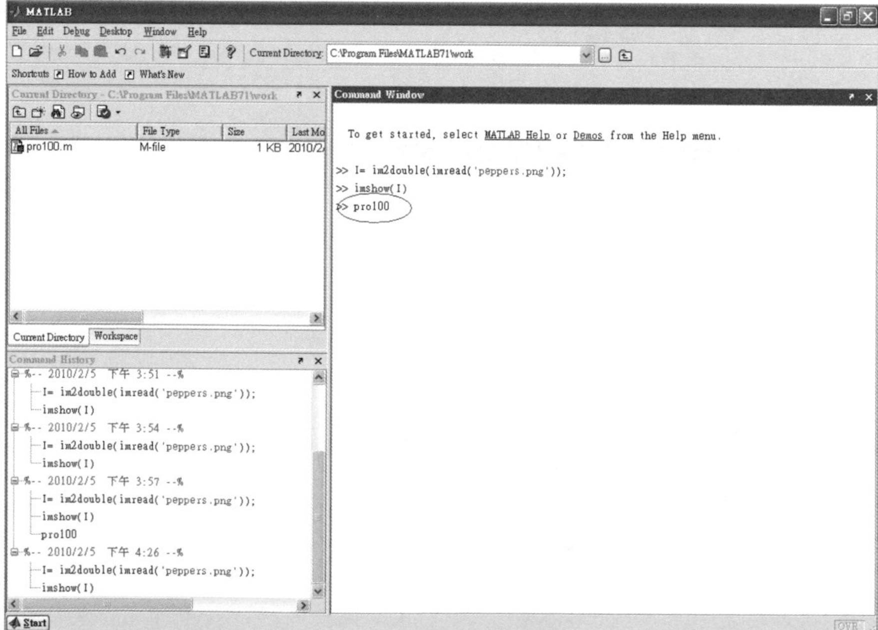

Step 9 執行結果與先前使用直譯器的結果相同

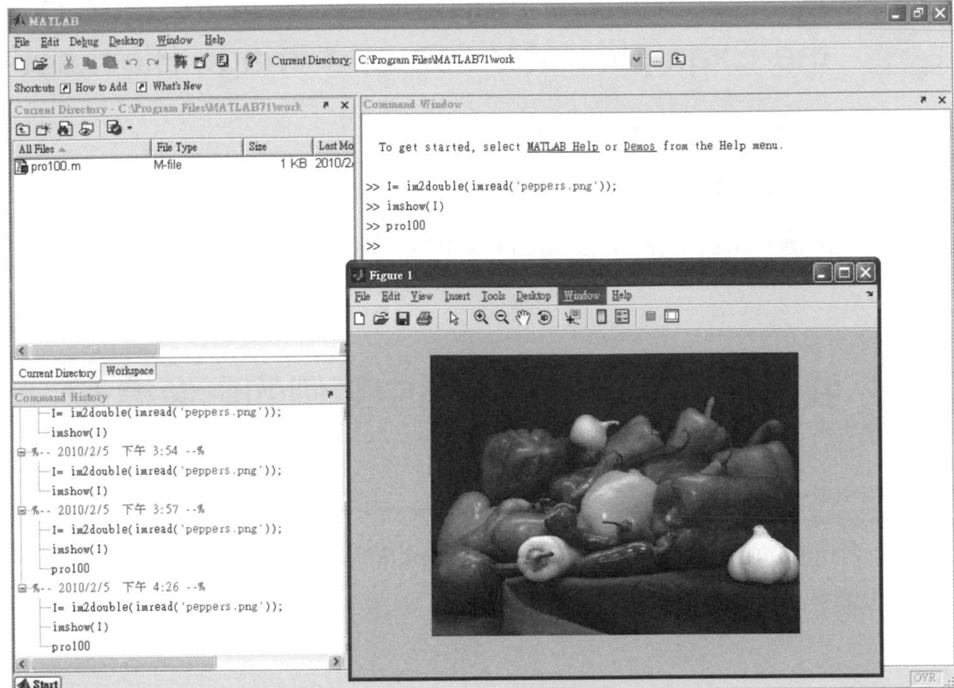

Step 10 加油!

0-3 Where Can We Find the Default Images in the Examples of Video and Image Blockset?

1. The path to the default images:

   C:\Program Files\MATLAB\toolbox\vipblks\vipdemos

2. The images files were appended CR-ROM in this book.

0-3 我們能在哪裡找到本書中範例的影像檔?

1. MatLab 預設的路徑：

   C:\Program Files\MATLAB\toolbox\vipblks\vipdemos

2. 本書所附的 CD ROM。

Import and View the AVI Files
輸入與觀看 AVI 檔案

1-1 Import and View the AVI Files
輸入並觀看 AVI 檔案

1-2 Export the AVI Files
輸出 AVI 檔案

1-3 Import and View RGB Signals in AVI File
輸入並觀看 AVI 檔案的 RGB 訊號

1-4 Export Multimedia Files
輸出到多媒體檔案

1-1 Import and View the AVI Files

A. SimuLink Layout

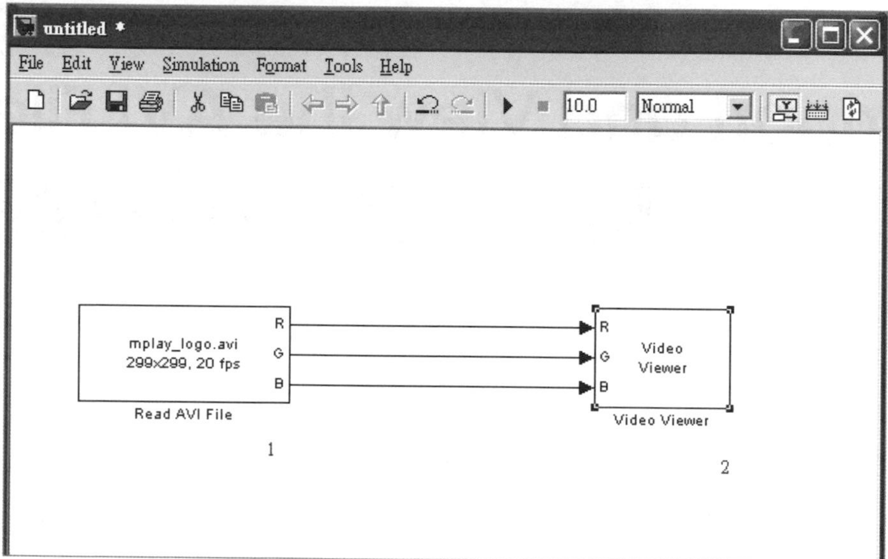

B. Path of Block

1. Video and Image Processing Blockset→Sources→Read AVI File
2. Video and Image Processing Blockset→Sinks→Video Viewer

C. Parameter Settings

(1)Read AVI File	(2)Video Viewer
Default	Default

D. Result

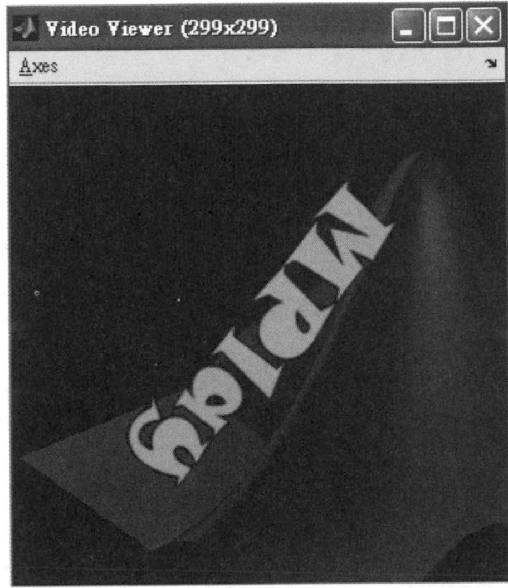

E. Translation

1. Default
2. View
3. Link
4. Video
5. Enhancement
6. Image
7. Process
8. File
9. Strengthen
10. Connect

F. Question

1. Keep running, and stop at the 120th second.

1-2 Export the AVI Files

A. SimuLink Layout

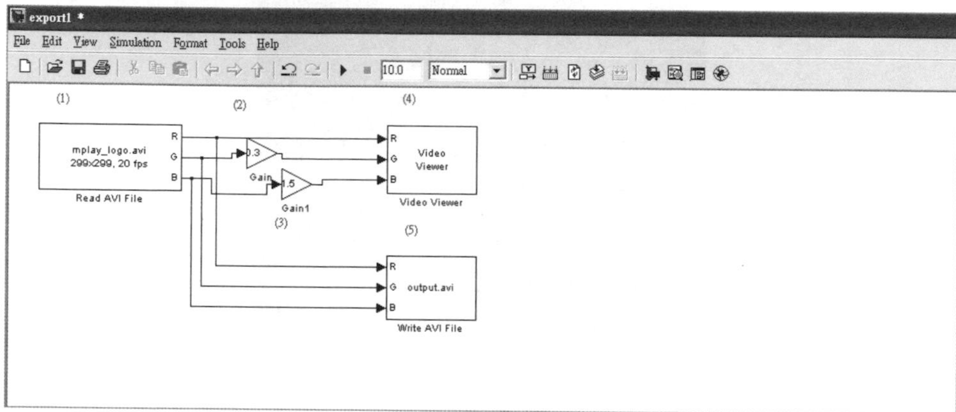

B. Path of the block

1. Video and Image Processing Blockset→Sources→Read AVI File
2. Simulink→Math Operations→Gain
3. Simulink→Math Operations→Gain
4. Video and Image Processing Blockset→Sinks→Video Viewer
5. Video and Image Processing Blockset→Sinks→Write AVI File

C. Parameter settings

(1)Read AVI File	(2)Gain	(3)Gain	(4)Video Viewer	(5)Write AVI File
default	Gain = 0.3	Gain = 1.5	default	default
	Output data type mode = Same as input	Output data type mode = Same as input		

D. Result

E. Translation

1. Parameter
2. Show
3. Create
4. Enable
5. Model
6. Section
7. Value
8. Set
9. Double
10. Play

F. Questions

1. What is the output, if the "Gain=0.5" in Gain block (2)? Why?
2. What is the output, if the "Gain=1.2" in Gain block (3)? Why?

1-3 Import and View RGB Signals in AVI File

A. SimuLink Layout

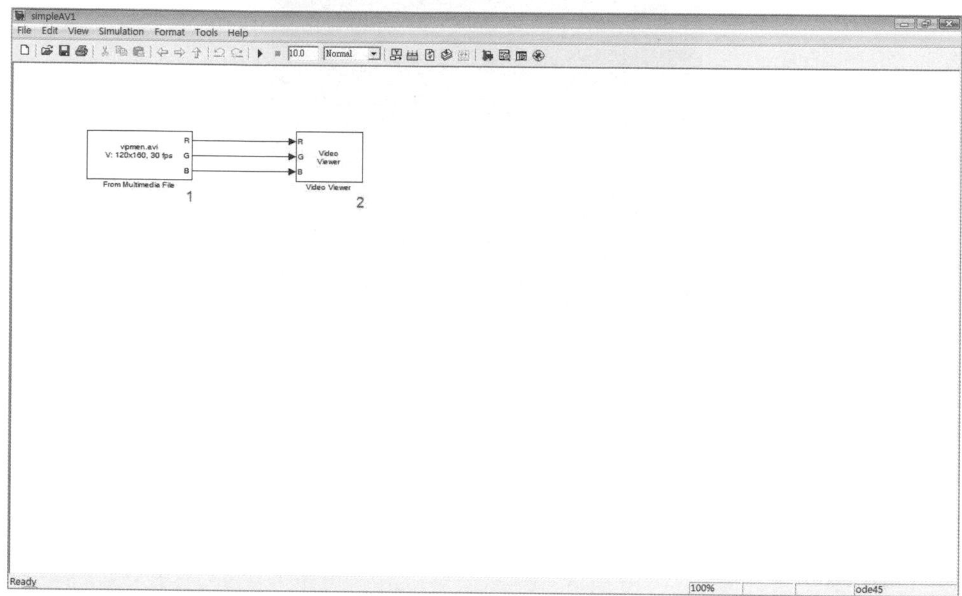

B. Path of the block

1. Video and Image Processing Blockset→Sources→From Multimedia file
2. Video and Image Processing Blockset→Sinks→Video Viewer

C. Parameter settings

(1)From Multimedia file	(2)Video Viewer
default	default

D. Result

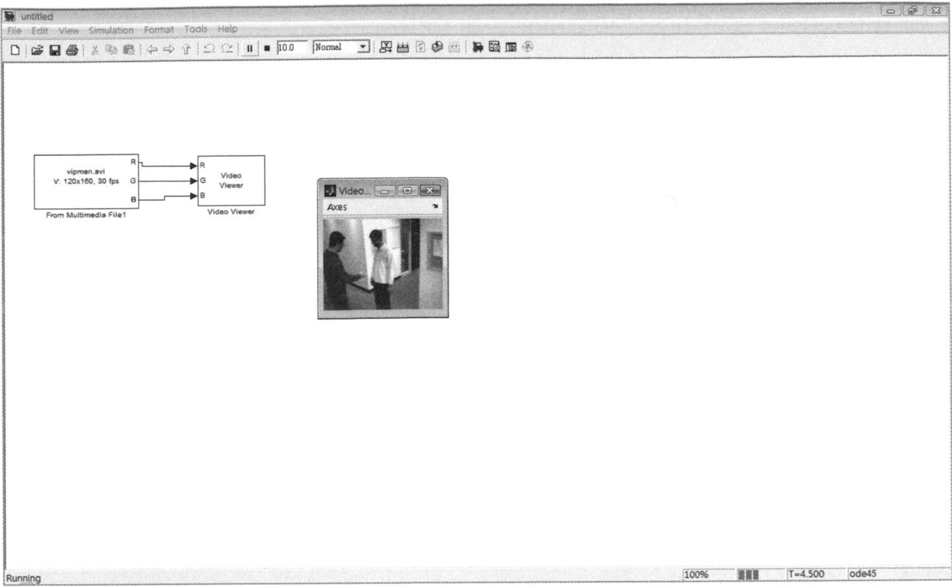

E. Translation

1. Multimedia
2. Standard
3. Audio
4. Viewer
5. Illustration
6. Inverter
7. Happen
8. Verify
9. Keep
10. Surround

F. Question

1. If the links is changed to connect as follows:

   (a)R→R G→G

   (b)R→R

   (c)R→G G→B B→R

   What will be displayed in the Video Viewer (2)?

Export Multimedia Files

A. SimuLink Layout

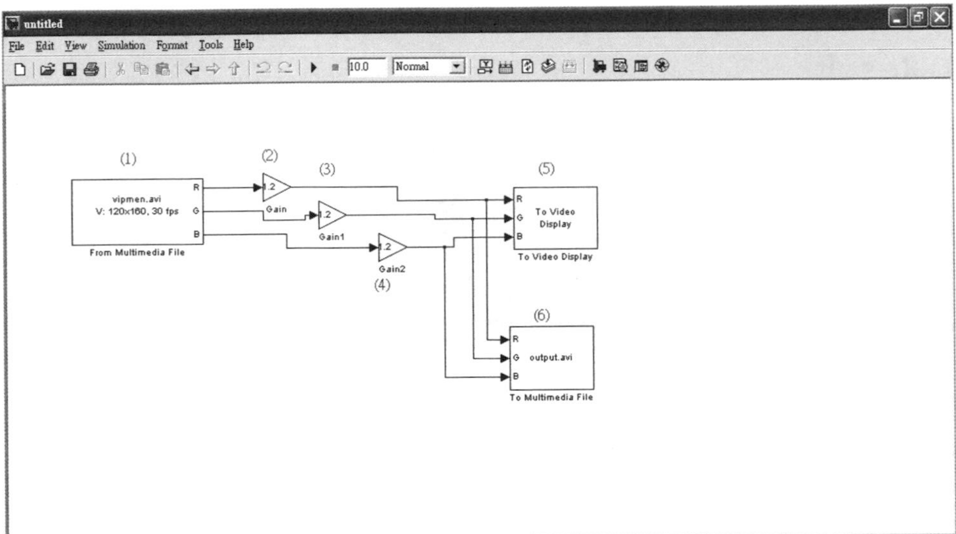

B. Path of Block

1. Video and Image Processing Blockset→Sources→From Multimedia File
2. Simulink→Math Operations→Gain
3. Simulink→Math Operations→Gain
4. Simulink→Math Operations→Gain
5. Video and Image Processing Blockset→Sinks→To Video Display
6. Video and Image Processing Blockset→Sinks→To Multimedia File

C. Parameter Settings

(1)From Multimedia File	(2)~(4)Gain	(5)To Video Display	(6)To Multimedia File
default	Gain = 1.2	default	Write = Video only
	Output data type mode = Same as input		

D. Result

E. Translation

1. Export
2. Figure
3. Procedure
4. Desktop
5. Assume
6. Equivalent
7. Already
8. Configuration
9. Setting
10. Directory

F. Question

1. If the 3 gains are changed to be 1.5 in block (2), 1.2 in block (3), and 1.5 in block (4), what will be the output?

Conversions
轉換

2-1 Converting Between Intensity and Binary Images Using Relational Operators
使用關係運算子來轉換灰階與二元值影像

2-2 Converting Between Intensity and Binary Images Using the Auto-threshold Block
使用自動臨界值法轉換灰階與二元值影像

2-3 Converting Color Information from R'G'B' into Gray Level
將彩色訊號轉為灰階

2-4 Chroma Resampling
色度重新取樣

2-1 Converting Between Intensity and Binary Images Using Relational Operators

A. Command Window

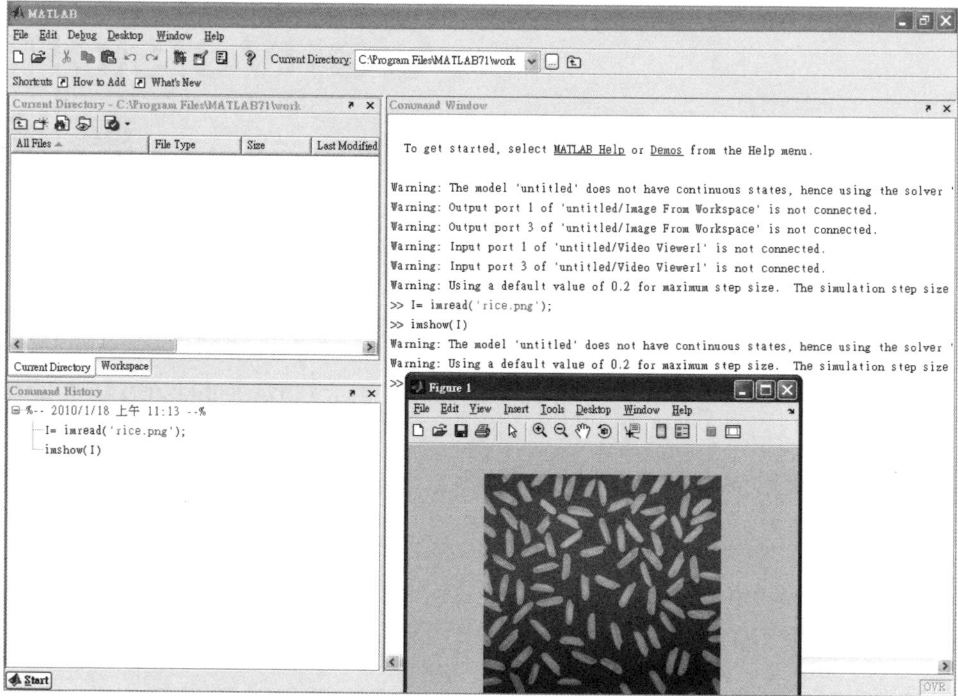


```
I= imread('rice.png');

imshow(I)
```

B. SimuLink Layout

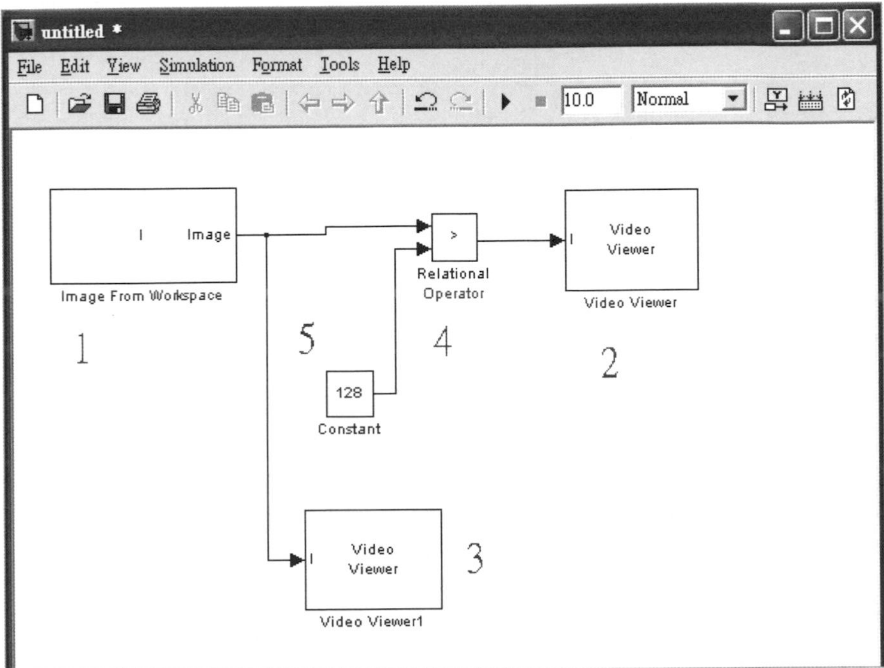

C. Path of Block

1. Video and Image Processing Blockset→Sources→Image From Workspace
2. Video and Image Processing Blockset→Sinks→Video Viewer
3. Video and Image Processing Blockset→ Sinks→Video Viewer
4. Simulink→Logic and Bit Operations→Relational Operator
5. Simulink→Sources→Constant

D. Parameter Settings

(1)Image From Workspace	(2)Constant	(3)Relational Operator	(4)Video Viewer	(5)Video Viewer
Value = I Output port labels = Image	Constant value=128	Relational Operator=>	Input image type =Intensity	Input image type =Intensity

E. Results

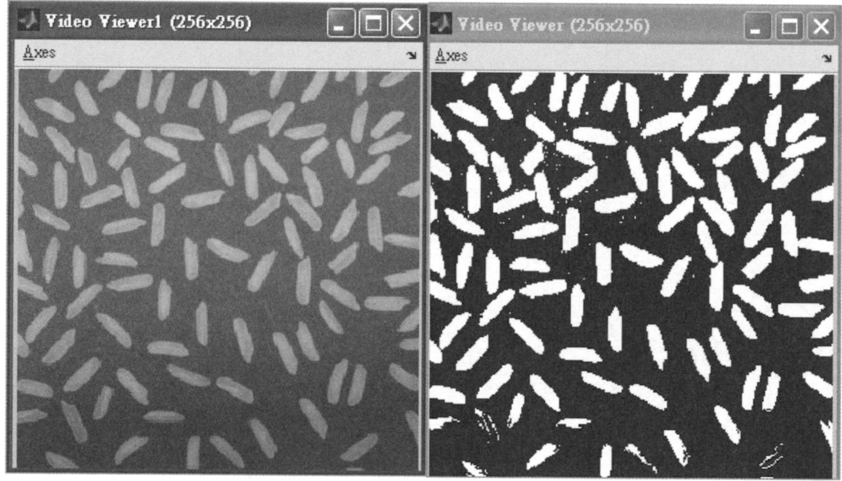

F. Translation

1. Present 顯示,展現
2. Watch
3. Image From Workspace
4. Relational
5. Operator

0 1 2 3 4 5 6 7 8 9 10

6. Constant
7. Source
8. Logic and Bit Operations
9. Input image type
10. Constant value

G. Question

1. Set “Constant value=150” in block (2), and what will be changed in the output image?

Converting Between Intensity and Binary Images Using the Auto-threshold Block

A. Command Window

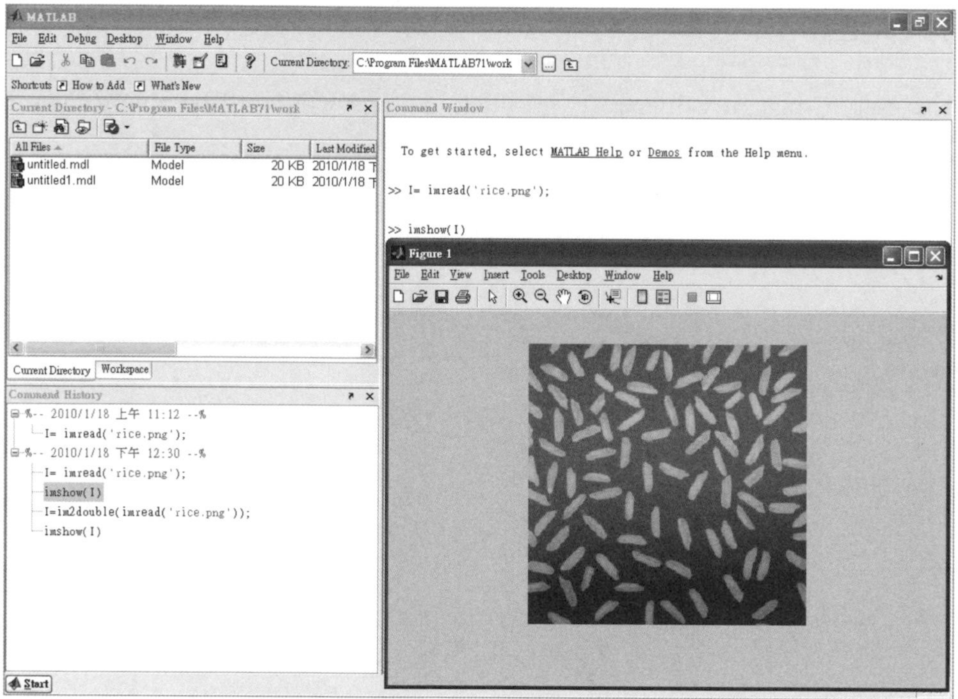


```
I= imread('rice.png');

imshow(I)
```

B. SimuLink Layout:

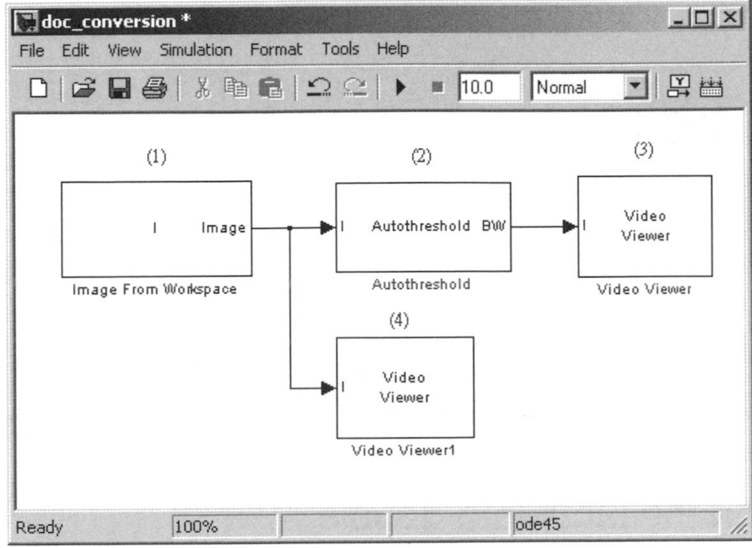

C. Path of Block

1. Video and Image Processing Blockset→Sources→Image From Workspace
2. Video and Image Processing Blockset→Conversions→Autothreshold
3. Video and Image Processing Blockset→Sources→Sinks→Video Viewer
4. Video and Image Processing Blockset→Sources→Sinks→Video Viewer

D. Parameter Settings

(1)Image From Workspace	(2)Autothreshold	(3)Video Viewer	(4)Video Viewer
Value: I	Thresholding operator: >	Input image type:Intensity	Input image type:Intensity
		Use colormap =ˇ	Use colormap =ˇ

E. Results

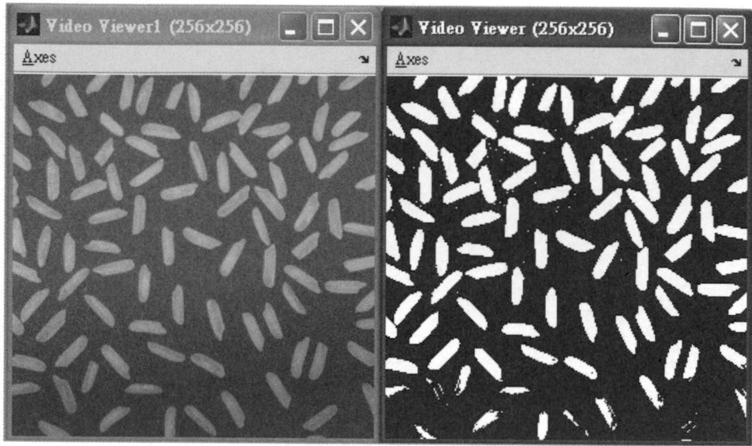

II. Convert the input into double

A. Comment Window

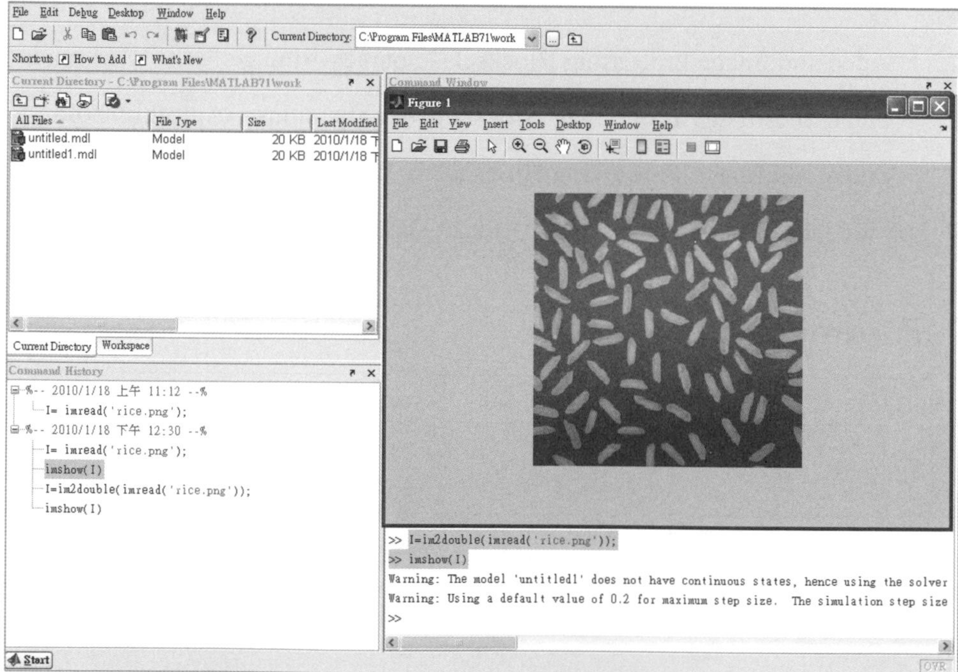

```
I=im2double(imread('rice.png'));

imshow(I)
```

B. Simulink Layout

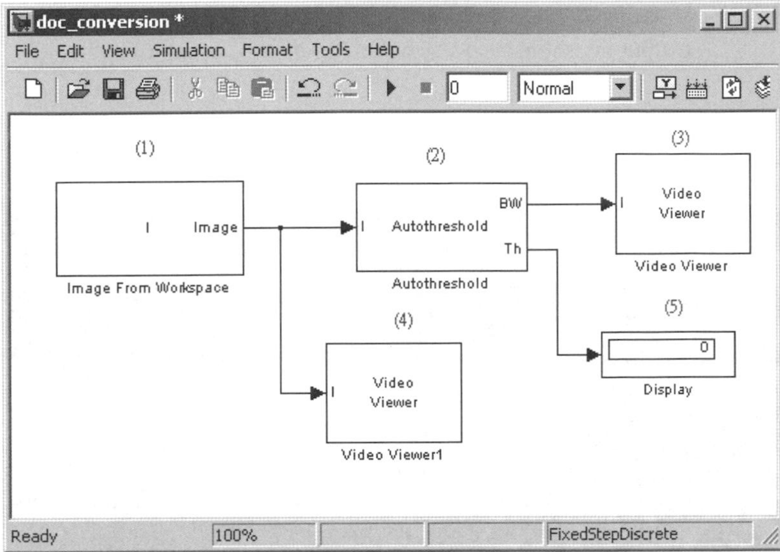

C. Path of Block

1. Video and Image Processing Blockset→Sources→Image From Workspace
2. Video and Image Processing Blockset→Conversions→Autothreshold
3. Video and Image Processing Blockset→Sources→Sinks→Video Viewer
4. Video and Image Processing Blockset→Sources→Sinks→Video Viewer
5. Simulink→Sinks→Display

D. Parameter Settings

(1)Image From Workspace	(2)Autothreshold	(3)Video Viewer	(4)Video Viewer	(5)Display
Value: I	Thresholding operator: >	Input image type:Intensity	Input image type:Intensity	default
	Output threshold =ˇ	Use colormap =ˇ	Use colormap =ˇ	

E. Results

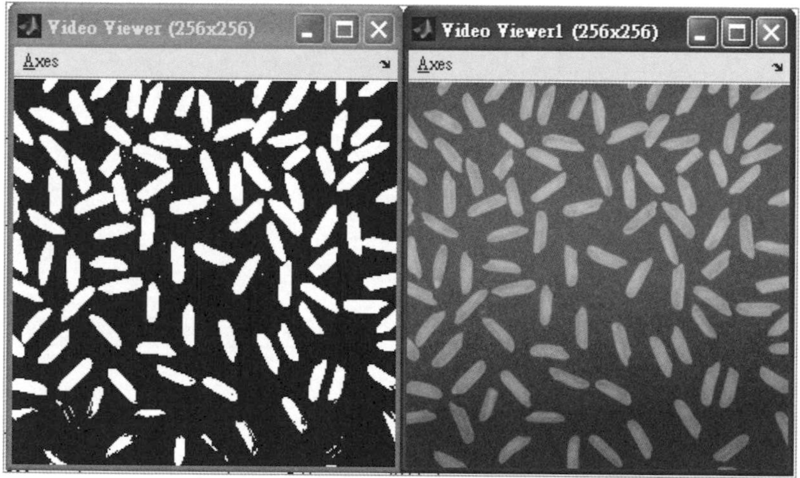

F. Translation

1. Rotate
2. Background
3. Fill
4. Gain
5. Geometric
6. Interpolation

7. Free
8. Neighbor
9. Center
10. Conner

G. Question

1. Select a new picture, and run the program again.

2-3 Converting Color Information from R'G'B' into Gray Level

A. Command Window

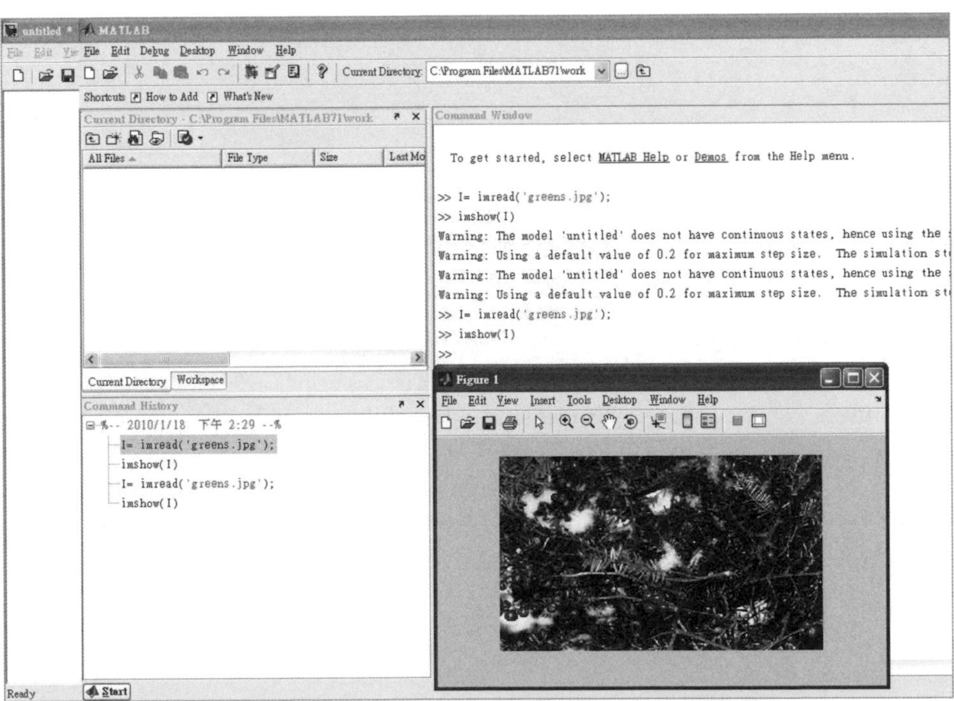

```
I= imread('greens.jpg');
imshow(I)
```

B. Simulink Layout

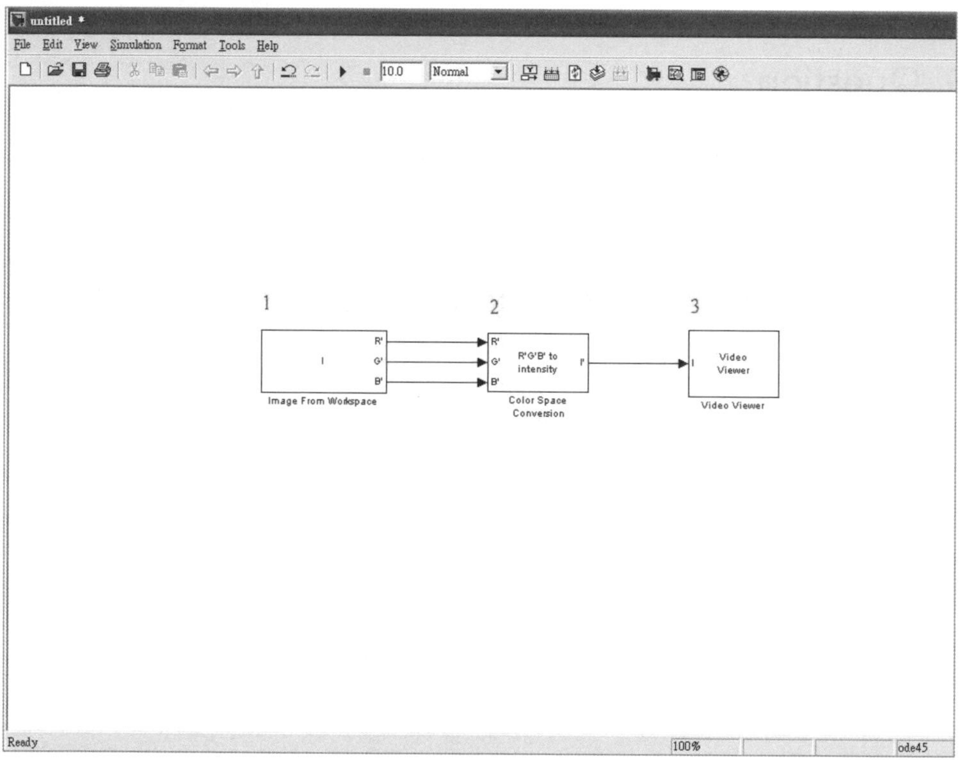

C. Path of Block

1. Video and Image Processing Blockset →Sources→Image From Workspace
2. Video and Image Processing Blockset→Conversions→Color Space Conversion
3. Video and Image Processing Blockset→Sinks→Video Viewer

D. Result

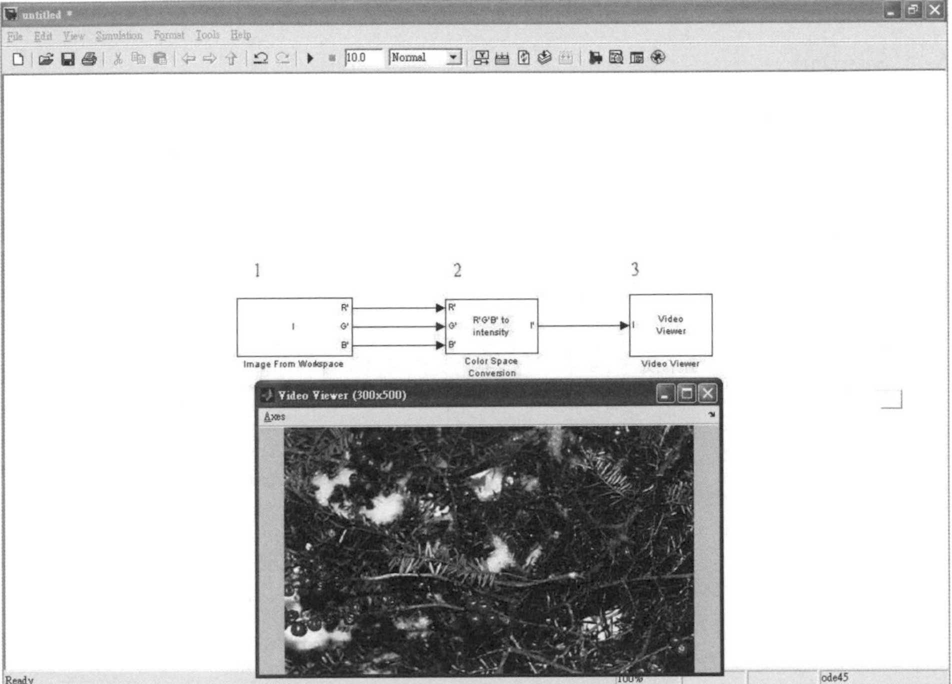

E. Parameter Settings

(1)Image From Workspace	(2)Color Space Conversion	(3)Video viewer
Value=I	Default	Input image type:= Intensity

F. Translation

1. Algorithm
2. Unsigned

3. Assemble
4. Display
5. Similar
6. Conversion
7. Variable
8. Command
9. Represent
10. Intensity

G. Question

1. Select a new picture, and run the program again.

Chroma Resampling

A. Command Window

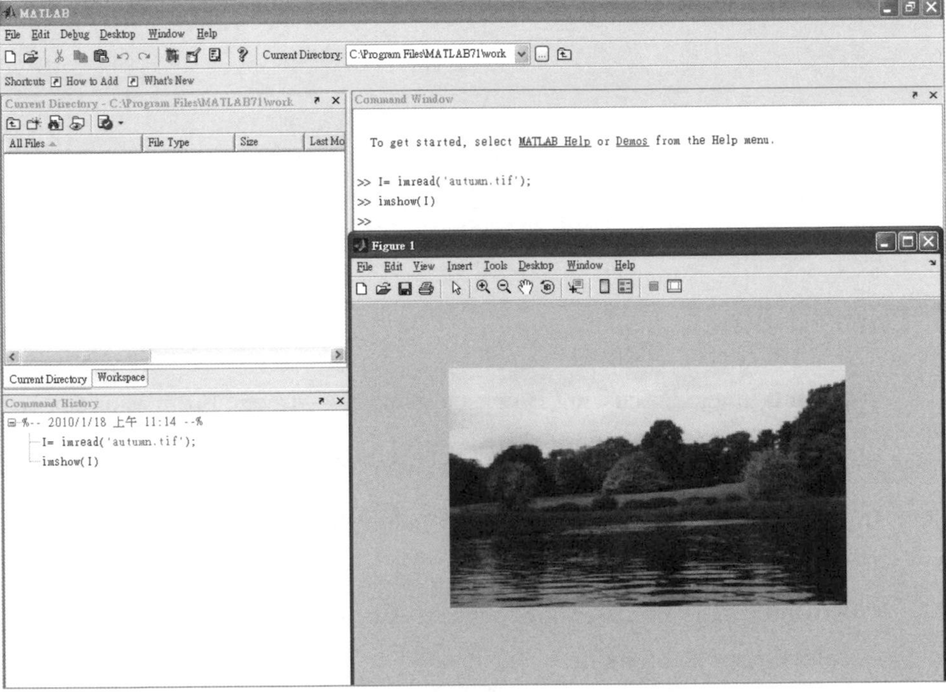


```
I= imread('autumn.tif');

imshow(I)
```

B. SimuLink Layout

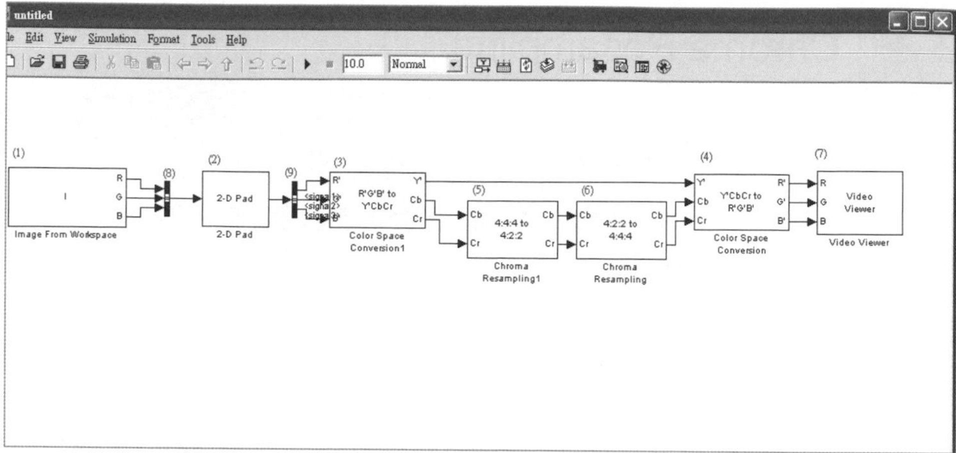

C. Path of Block

1. Video and Image Processing Blockset→sources→Image From Workspace
2. Video and Image Processing Blockset→Utiltites→2-D Pad
3. Video and Image Processing Blockset→Conversions →Color Space Conversion
4. Video and Image Processing Blockset→Conversions →Color Space Conversion
5. Video and Image Processing Blockset→Conversions →Chroma Resampling
6. Video and Image Processing Blockset→Conversions →Chroma Resampling
7. Video and Image Processing Blockset→Sinks→Video Viewer
8. Simulink→Signal Routing→Bus Creator
9. Simulink→Signal Routing→Bus Selector

D. Parameter Settings

(1)Image From Workspace	(2)2-D Pad	(3)Color Space Conversion	(4)Color Space Conversion
Value = I	Method=Symmetric Pad rows at = Right Pad size along rows = 1 Pad columns at = No padding	Default	Conversion=Y'CbCr to R'G'B'.

(5)Chroma Resampling	(6)Chroma Resampling	(7)Video Viewer	(8)Bus Creator	(9)Bus Selector
Resampling= 4:4:4to4:2:2	Resampling= 4:2:2to4:4:4	Select signal 3	Number of inputs = 3	Select signal 3

E. Results

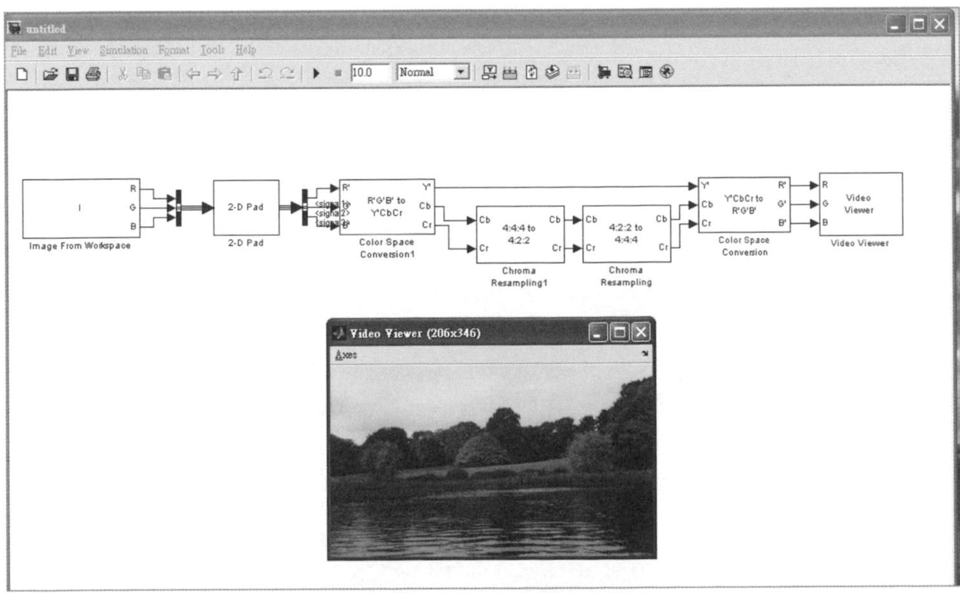

F. Translation

1. Component
2. Mother board
3. Following
4. Bus
5. Require
6. Instruction
7. Crack
8. Chroma
9. Sample
10. Transmitter

G. Question

1. Please select a new picture, and run the program again.

Geometric Transformation 幾何轉換

3-1 Rotation of Image
影像旋轉

3-2 Resizing an Image
重新調整影像大小

3-3 Cropping the Image
剪輯影像

3-1 Rotation of Image

A. SimuLink Layout

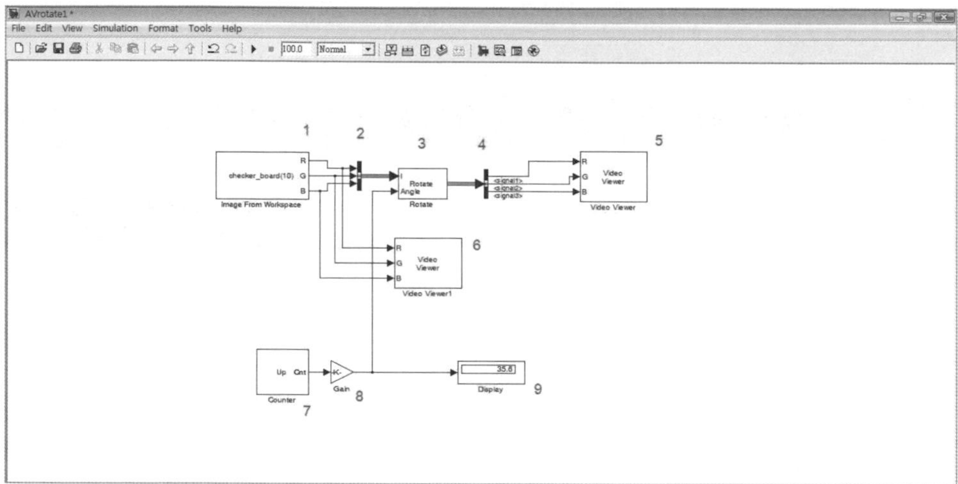

B. Path of Block

1. Video and Image Processing Blockset→Sources→Image From Workspace
2. Simulink→Signal Routing→Bus Creator
3. Video and Image Processing Blockset→Geometric Transformations →Rotate
4. Simulink→Signal Routing→Bus Selector
5. Video and Image Processing Blockset→Sinks→Video Viewer
6. Video and Image Processing Blockset→Sinks→Video Viewer
7. Signal Processing Blockset→Signal Management→Switches and Counters →Counter
8. Simulink→Math Operations→Gain
9. Signal Processing Blockset→Signal Processing Sinks→Display

C. Parameter Settings

(1)Image From Workspace	(2)Bus Creator	(3)Rotate	(4)Bus Selector
default	Number of input=3	Rotation angle source=input port	Select signal 3
		Maximum angle=pi	
		Display rotated image in=center	
		Sine value computation method =Trigonometric function	
		Background fill value=0	
		Interpolation method=Bilinear	

(5)Video Viewer	(6)Video Viewer	(7)Counter	(8)Gain	(9)Display
default	default	Count event = Free running	Gain=pi/180	default
		Counter size = 16 bits		
		Output = Count		
		Clear the Reset input check box.		
		Sample time = 1/30		

D. Results

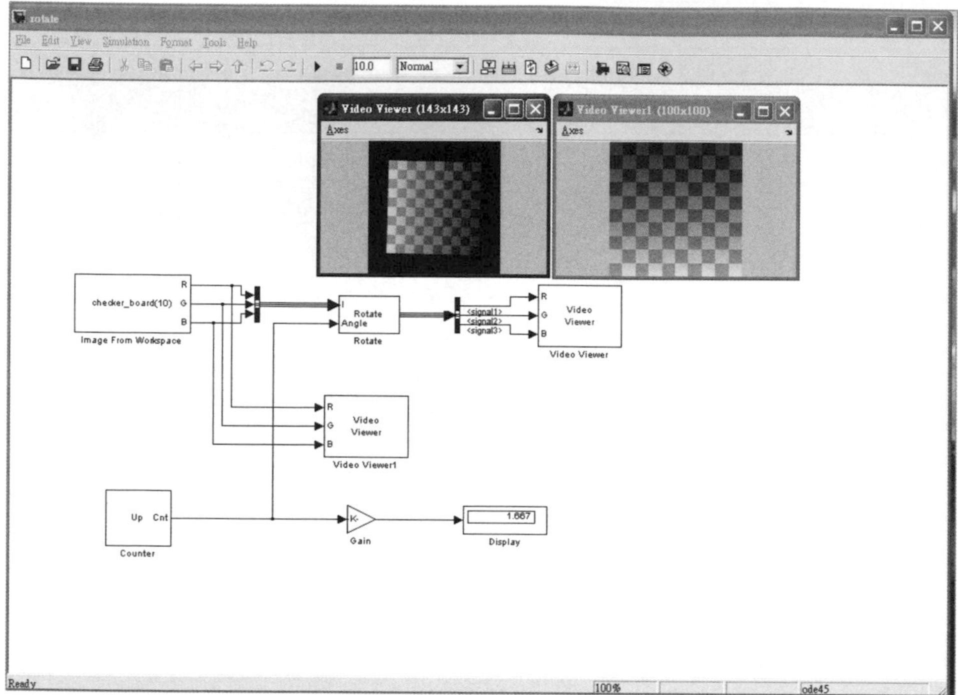

E. Translation

1. Component
2. Mother board
3. Following
4. Bus
5. Require
6. Instruction
7. Crack
8. Chroma
9. Sample
10. Transmitter

F. Question

1. Change the parameters in the rotating Video Viewer (Block (3))?

   (a) Interpolation method=Bilinear→Bicubic→Nearest neighbor

   (b) Display rotated image in=center→Top-left Conner

3-2 Resizing an Image

A. Comments in Workspace

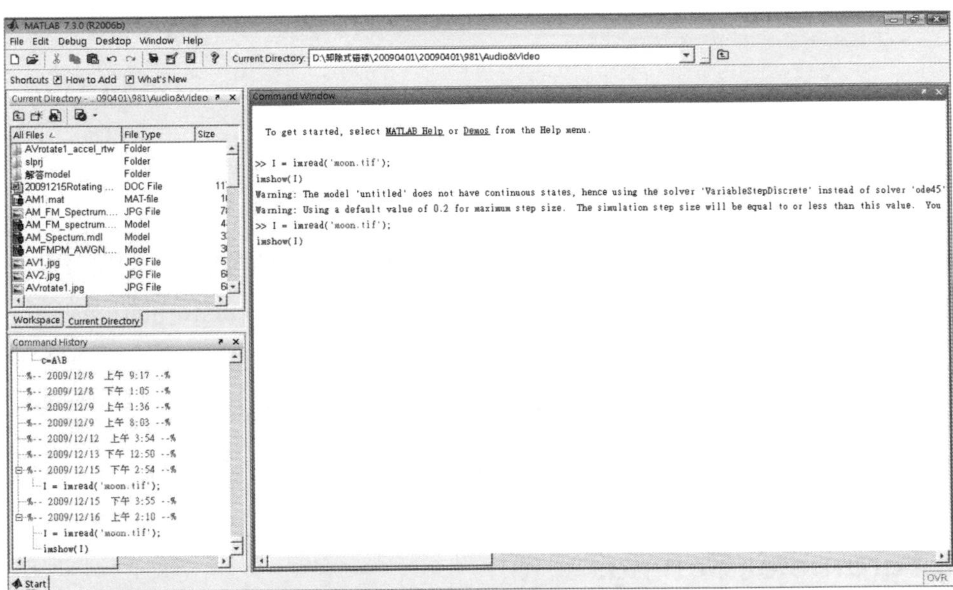


```
I=imread('moon.tif');
Imshow(I)
```

B. SimuLink Layout

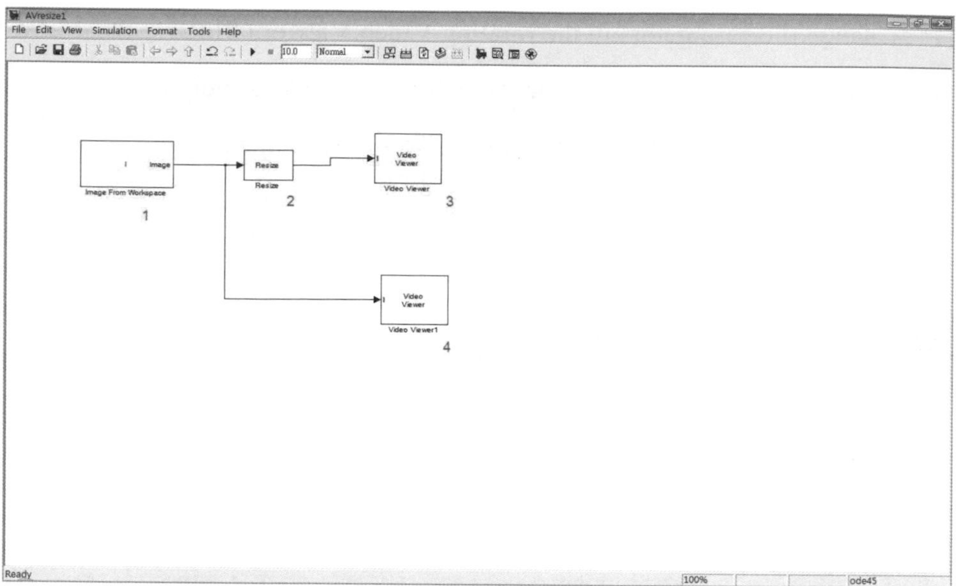

C. Path of Block

1. Video and Image Processing Blockset→Sources→Image From Workspace
2. Video and Image Processing Blockset→Geometric Transformations →Resize
3. Video and Image Processing Blockset→Sinks→Video Viewer
4. Video and Image Processing Blockset→Sinks→Video Viewer

D. Parameter Settings

(1)Image From Workspace	(2)Resize	(3)Video Viewer	(4)Video Viewer
Value=I	Resize factor in %=50	Input image type=Intensity	Input image type=Intensity
Output port labels=Image			

E. Result

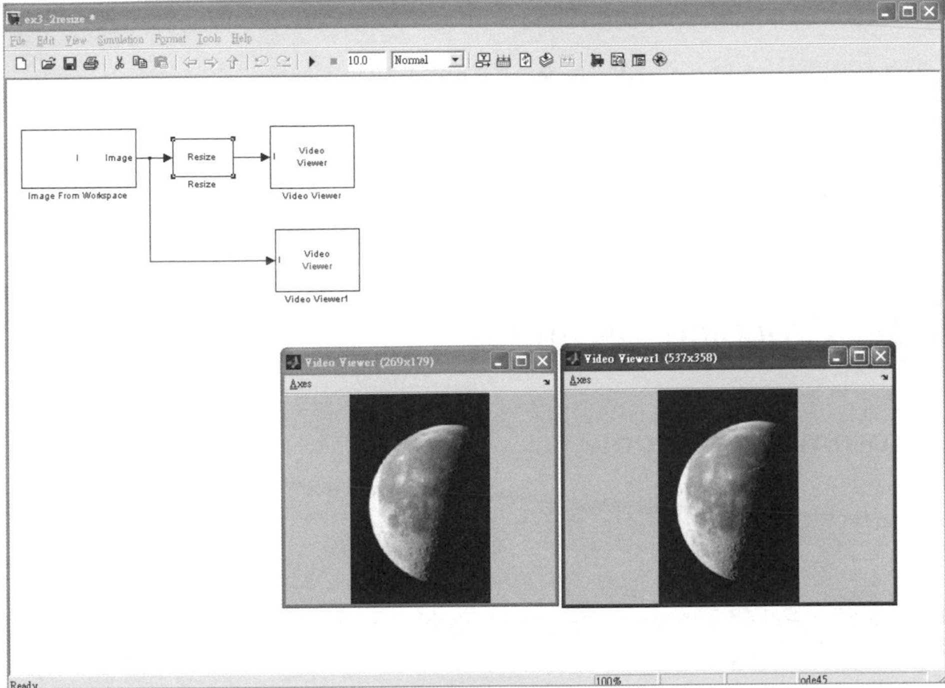

F. Translation

1. Resize
2. Factor
3. Label
4. Strong
5. Port
6. Charge
7. Pixel
8. Shape
9. Rectangular
10. Circular

G. Question:

1. What will be happened in the resiting Video Viewer? Why?

   (a) Resizing factor=10   in Resize block (2)

   (b) Resizing factor=30

   (c) Resizing factor=80

3-3 Cropping the Image

A. Command window

```
I = imread('coins.png');
imshow(I)
```

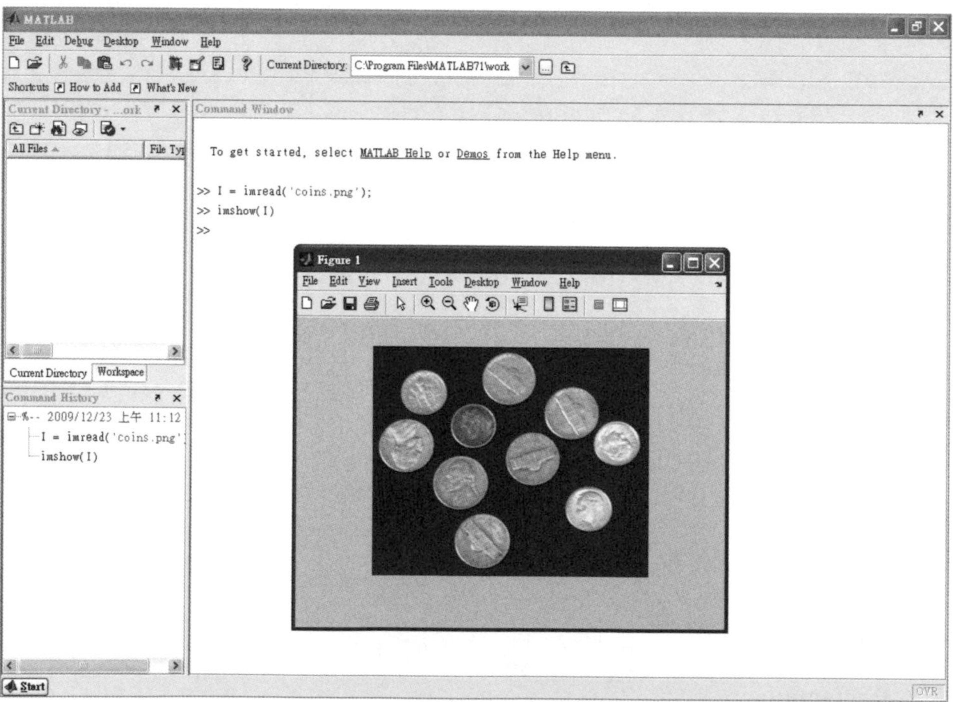

B. SimuLink Layout

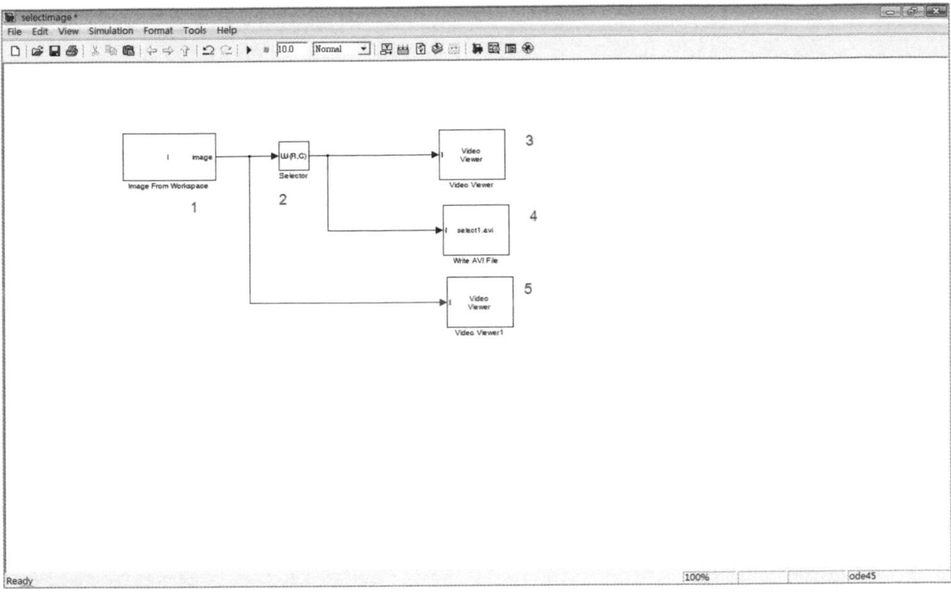

C. Path of Block

1. Video and Image Processing Blockset→Sources→Image From Workspace
2. Simulink→Signal Routing→Selector
3. Video and Image Processing Blockset→Sinks→Video Viewer
4. Video and Image Processing Blockset→Sinks→Write AVI file
5. Video and Image Processing Blockset→Sinks→Video Viewer

D. Parameter Settings

(1)Image From Workspace	(2) Selector	(3)Video Viewer	(4)Write AVI file	(5)Video Viewer
Value = I	Input type = Matrix	Input image type = Intensity	File name =select1.avi	Input image type = Intensity
Output port labels = Image	Rows = 140			
	Columns = 200			
	Select the Use index as starting value check Box			
	Output port dimensions = [70 70]			

E. Results

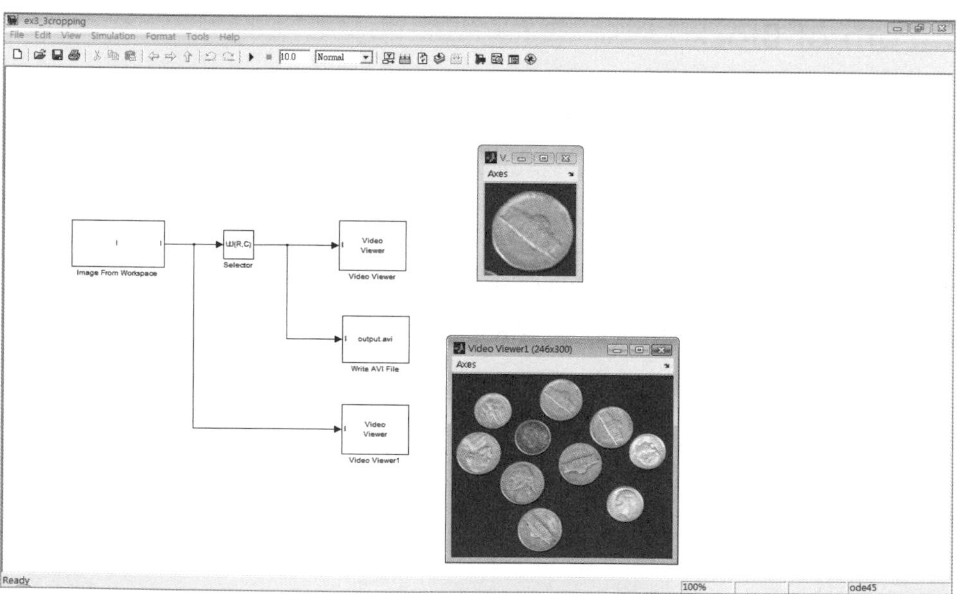

F. Translation

1. Crop
2. Selector
3. Dimension
4. Start
5. Check box
6. Stand
7. Icon
8. Busy
9. Exist
10. Current directory

G. Questions

1. What is the output, if the settings are changed as follows? Why?

   (a) Rows = 140→1; Columns = 200→115;

   Output port dimensions = [65 65]

   (b) Rows = 90; Columns = 140;

   Output port dimensions = [65 65]

2. Please crop a coin.

筆記頁

Morphological Operations
形態學上的操作

4-1 Object Count in an Image
在影像中計數物件個數

4-2 Correct the Non-uniform Illumination
修正非均勻照度

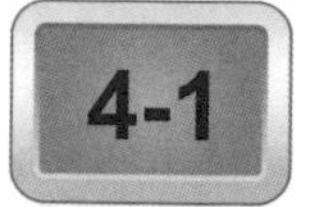

Object Count in an Image

A. Command Window

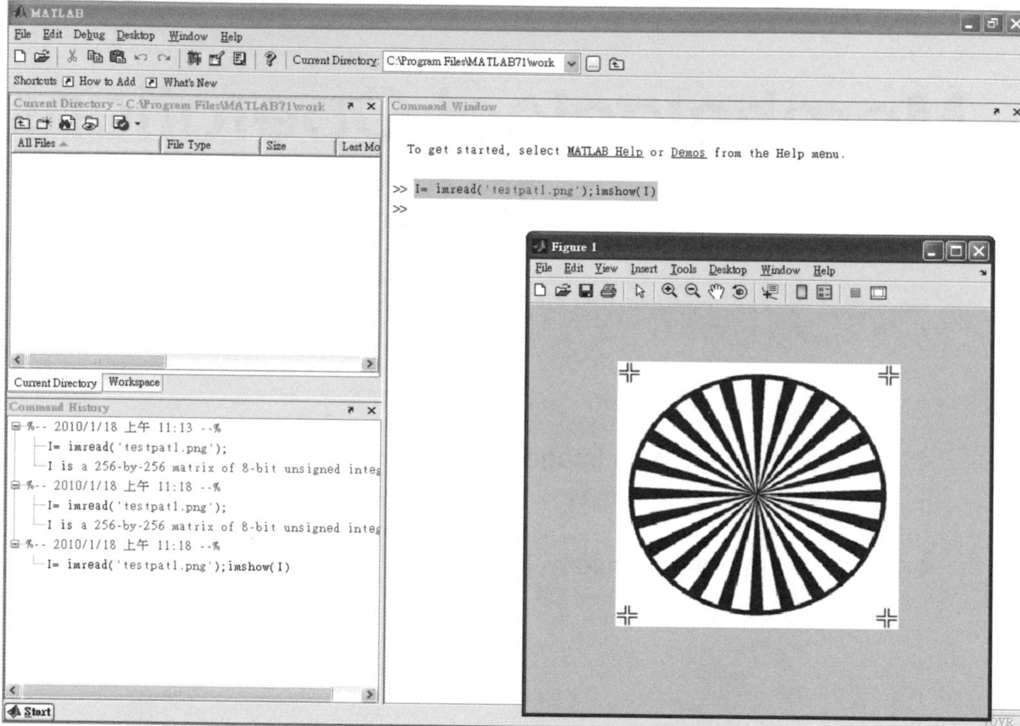


```
I= imread('testpat1.png');

imshow(I)
```

B. SimuLink Layout

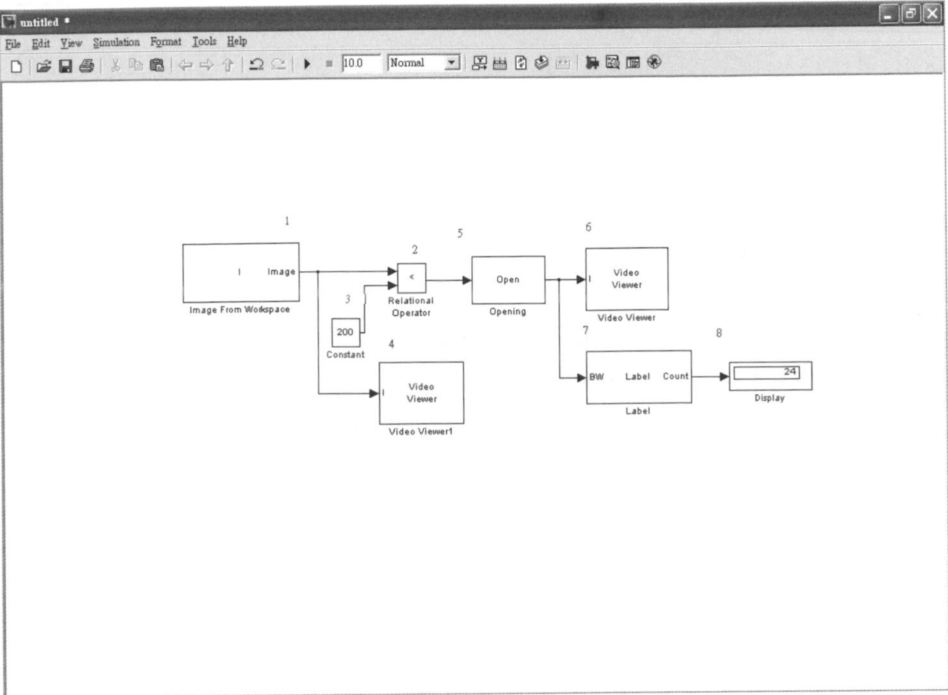

C. Path of Block

1. Video and Image Processing Blockset→Sources→Image From Workspace
2. Simulink→Logic and Bit Operations→Relational Operator
3. Simulink→Sources→Constant
4. Video and Image Processing Blockset→Sinks→Video Viewer
5. Video and Image Processing Blockset→Morphological Operations →Opening
6. Video and Image Processing Blockset→Sinks→Video Viewer
7. Video and Image Processing Blockset→Morphological Operations→Label
8. Signal Processing Blockset→Signal Processing Sinks→Display

D. Parameter Settings

(1)Image From Workspace	(2)Relational Operator	(3)Constant	(4)Video Viewer
Value = I	Relational Operator= <	Constant value =200	Input image type parameter = Intensity.
Output port labels = Image			

(5)Opening	(6)Video Viewer	(7)Label	(8)Display
Default	Input image type parameter = Intensity.	Connectivity = 8	Default
		Output=Number of labels	

E. Results

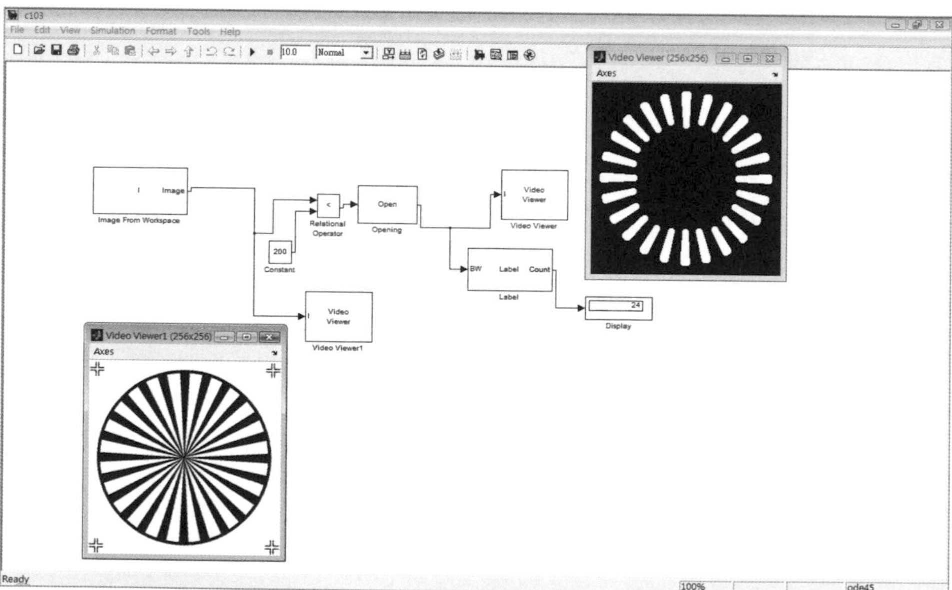

F. Translation

1. Hit
2. Speed
3. Display
4. Resistor
5. Structure
6. Operator
7. Architecture
8. Operation system
9. Memory
10. Indicate

G. Question

1. Set "Constant value =250" in block (3), and find the changes in the output.

4-2 Correct the Non-uniform Illumination

A. Command Window

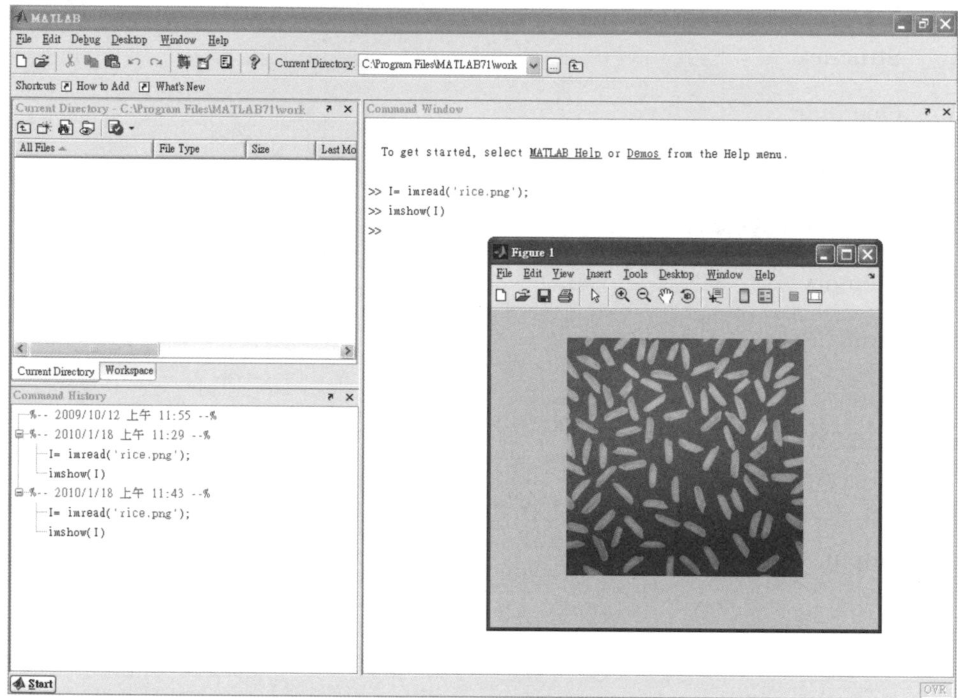


```
I= imread('rice.png');

imshow(I)
```

B. SimuLink Layout

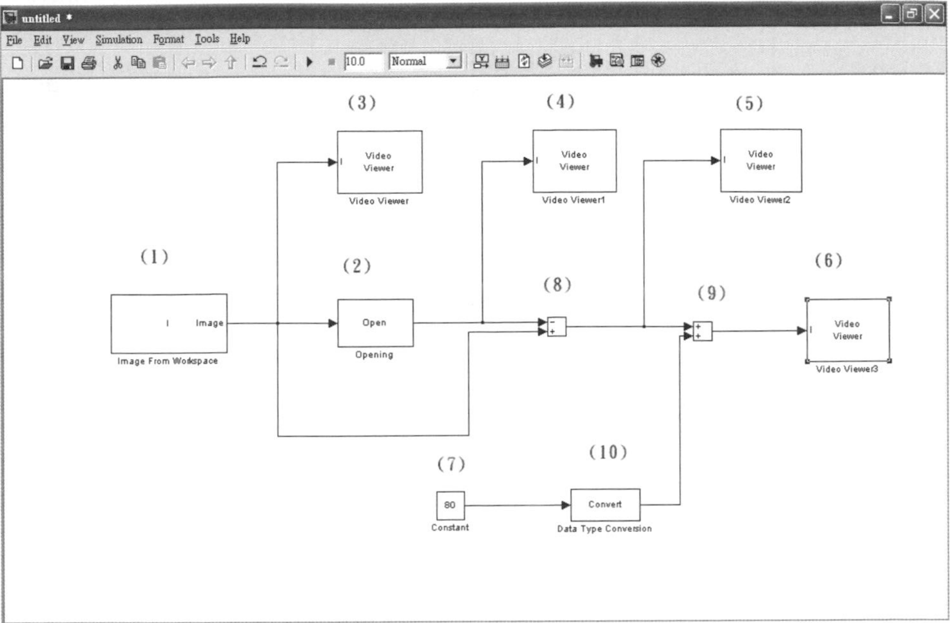

C. Path of Block

1. Video and Image Processing Blockset→Sources→Image From Workspace
2. Video and Image Processing Blockset→Morphological Operations→ Opening
3. Video and Image Processing Blockset→Sinks→Video Viewer
4. Video and Image Processing Blockset→Sinks→Video Viewer
5. Video and Image Processing Blockset→Sinks→Video Viewer
6. Video and Image Processing Blockset→Sinks→Video Viewer
7. Simulink→Sources→Constant
8. Simulink→Math Operations→Sum
9. Simulink→Math Operations→Sum
10. Simulink→Signal Attributes→Data Type Conversion

D. Parameter Settings

(1)Image From Workspace	(2)Opening	(3)~(6) Video Viewer	(7)Constant	(8)~(9)Sum	(10)Data Type Conversion
Value = R	Neighborhood or structuring element sourse =Specify via dialog	Input image type =Intensity	Constant value =80	Icon shape = rectangular	Output dimensionality =Column vector.
Output port labels = Image	Neighborhood or structuring element =strel('disk',20)			List of signs = -+	
				List of signs = ++	

E. Results

- Viewer(3)

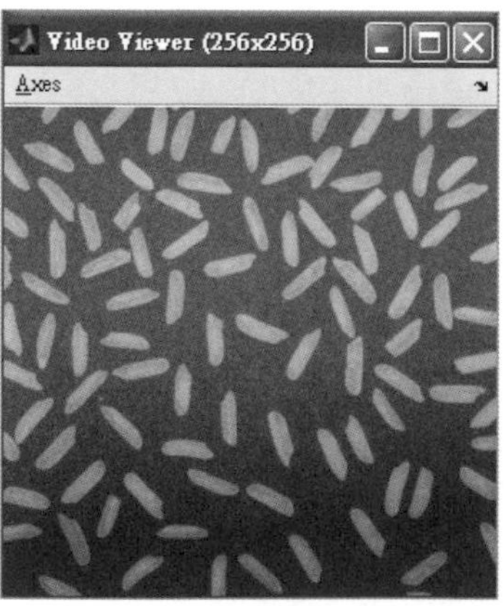

- Viewer(4)

- Viewer2(5)

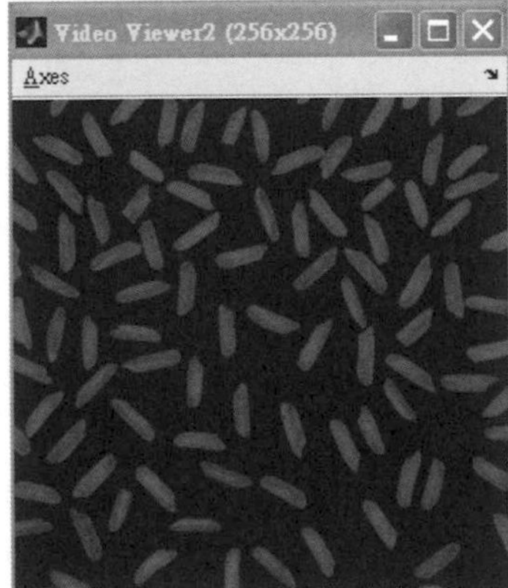

- Viewer3(6)

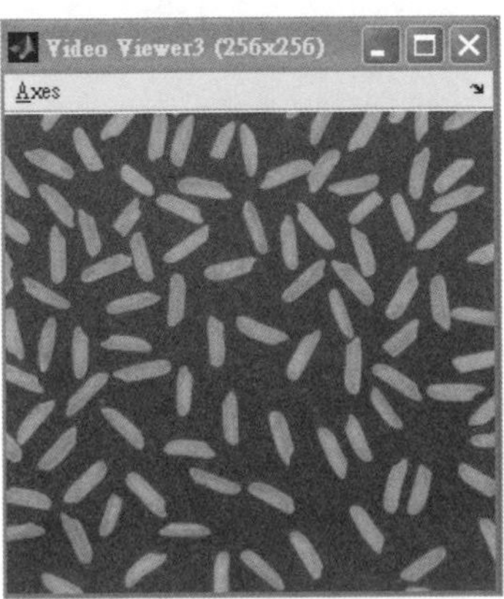

F. Translation

1. Point out
2. Demonstration
3. Example
4. Remember
5. Transistor
6. Amplifier
7. Block
8. Model
9. Graphic
10. Interrupt

G. Questions

1. Set “Constant value = 10, 50, and 70” in block (7), what will be changed in the output image?
2. Set “Neighborhood or structuring element =strel('disk',30)” in block (2), how will the output images be changed?
3. Select color map in the viewers of blocks (3) to (6).

Feature Extraction
特徵提取

5-1 Edge Detection
偵測圖像的邊緣

5-2 Line Detection
偵測影像中的線

5-1 Edge Detection

A. Command Windows

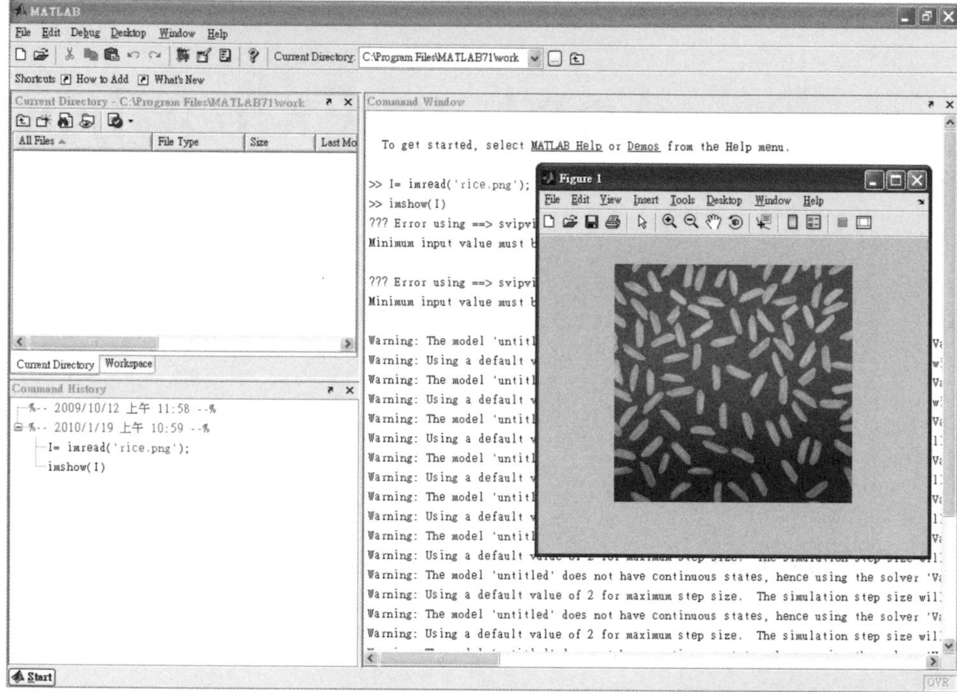


```
I= imread('rice.png');
imshow(I)
```

B. Simulink Layout

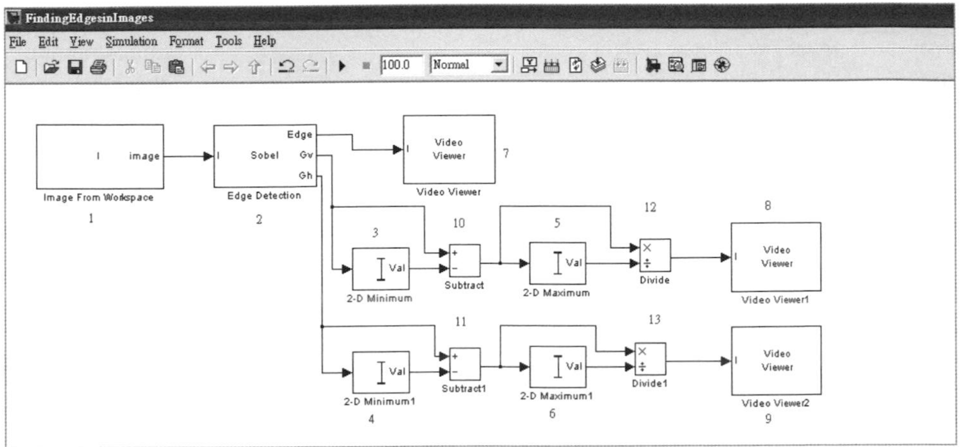

C. Path of Block

1. Video and Image Processing Blockset→Sources→Image From Workspace
2. Video and Image Processing Blockset→Analysis & Enhancement →Edge Detection
3. Video and Image Processing Blockset→Statistics→2-D Minimum
4. Video and Image Processing Blockset→Statistics→2-D Minimum
5. Video and Image Processing Blockset→Statistics→2-D Maximum
6. Video and Image Processing Blockset→Statistics→2-D Maximum
7. Video and Image Processing Blockset→Sinks→Video Viewer
8. Video and Image Processing Blockset→Sinks→Video Viewer
9. Video and Image Processing Blockset→Sinks→Video Viewer
10. Simulink→Math Operations→Subtract
11. Simulink→Math Operations→ Subtract
12. Simulink→Math Operations→Divide
13. Simulink→Math Operations→Divide

D. Parameter Settings

(1)Im from Workspace Main pane	(2)Im from Workspace Main pane	(3)Data Types pane	(4)Output type	(5)Select the Edge thinning check box
Value = I	Output port labels = Image	Output data type = double	Output type = Binary image and gradient components	Select the Edge thinning check box

(6)~(8)Video viewer	(9)~(10)Divide	(11)~(12)Subtract	(13)Minimum	(14)Maximum
default	default	default	Mode=Value	Mode=Value

E. Results

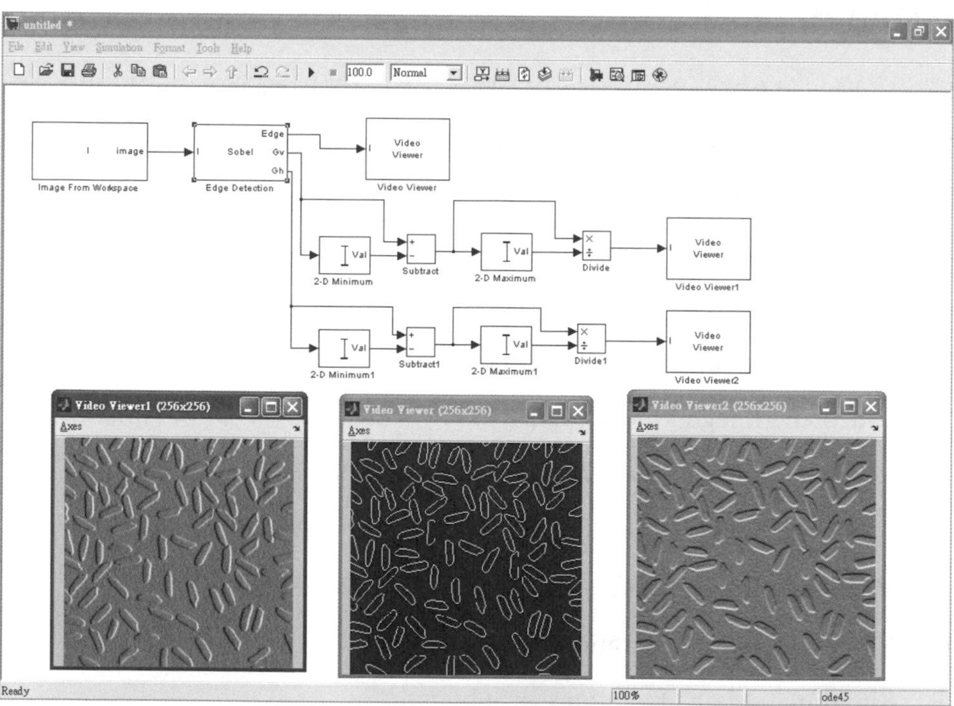

F. Translation

1. Add
2. Subtract
3. Review
4. Minimum
5. Maximum
6. Divide
7. Edge thinning
8. Detection
9. Multiple
10. Integer

G. Question

1. What will be changed in the output, if we clear the check box of "Main Edge Thinning" in Sobel block (2)?

5-2 Line Detection

A. Command Window

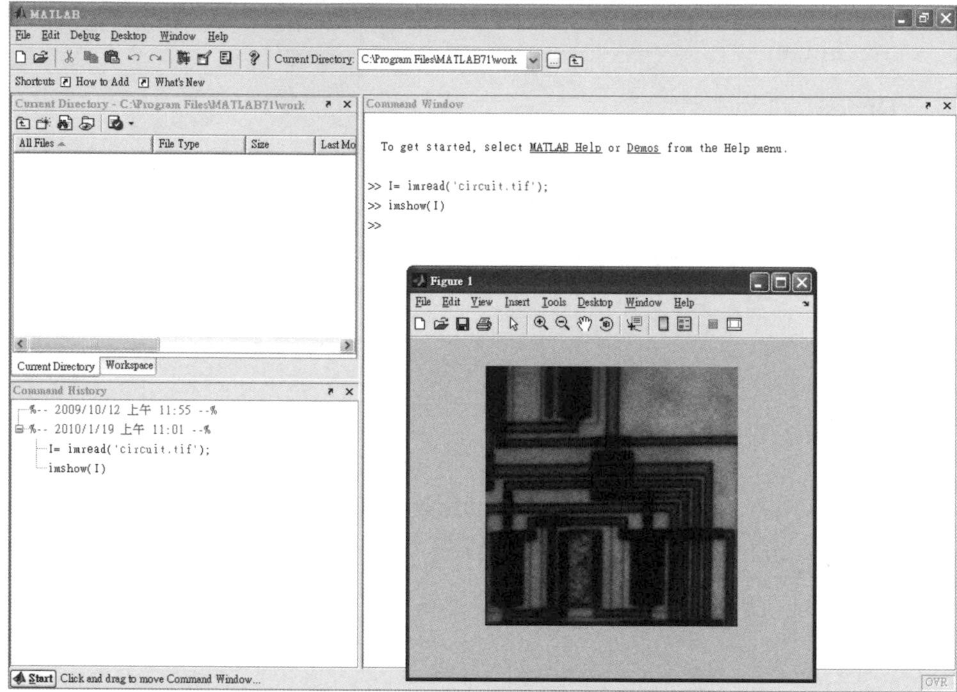


```
I= imread('circuit.tif');

imshow(I)
```

B. SimuLink Layout

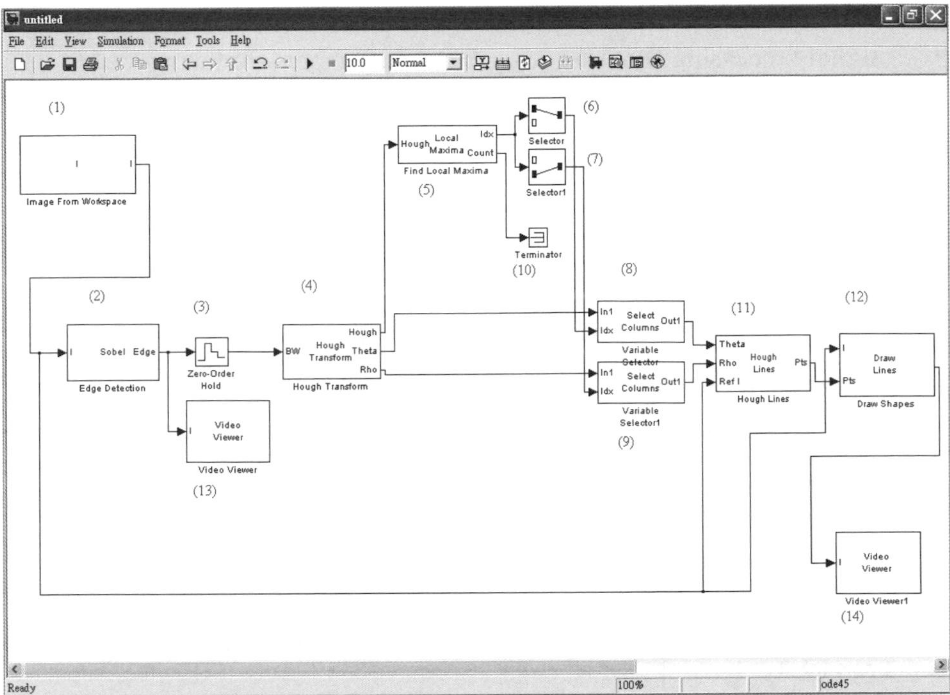

C. Path of Block

1. Video and Image Processing Blockset→Sources→Image From Workspace
2. Video and Image Processing Blockset→Analysis & Enhancement →Edge Detection
3. Simulink→Discrete→Zero-order Hold
4. Video and Image Processing Blockset→Transforms→Hough Transform
5. Video and Image Processing Blockset→Statistics→Find Local Maxima
6. Simulink→Signal Routing→Selector
7. Simulink→Signal Routing→Selector

8. Signal Processing Blockset→Signal Management→Indexing →Variable Selector
9. Signal Processing Blockset→Signal Management→Indexing →Variable Selector
10. Simulink→Sinks→Terminator
11. Video and Image Processing Blockset→Transforms→Hough Lines
12. Video and Image Processing Blockset→Text & Graphics→Draw Shapes
13. Video and Image Processing Blockset→Sinks→Video Viewer
14. Video and Image Processing Blockset→Sinks→Video Viewer

D. Parameter Settings

(1)mage From Workspace	(2)Edge Detection	(3)Zero-order Hold	(4)Hough Transform
Value = I	Default	Default	Theta resolution (radians) = pi/360
Output port labels = I			Select the Output theta and rho values check box.

(5)Find Local Maxima	(6)Selector	(7)Selector	(8)~(9)Variable Selector
Maximum number of local maxima (N) = 1	Index mode = Zero-based	Index mode = Zero-based	Select = Columns
Select the Input is Hough matrix spanning full theta range check box	Elements (-1 for all elements) = 0	Elements (-1 for all elements) = 1	Index mode = Zero-based
	Input port width = 2	Input port width = 2	

(10)Terminator	(11)Hough Lines	(12)Draw Shapes	(13)~(14)Video Viewer
Default	Sine value cmputation method =Trigomomeric function	Input image type =Intensity	Input image type =Intensity
		Shape = Lines	
		Border intensity = White	

E. Results

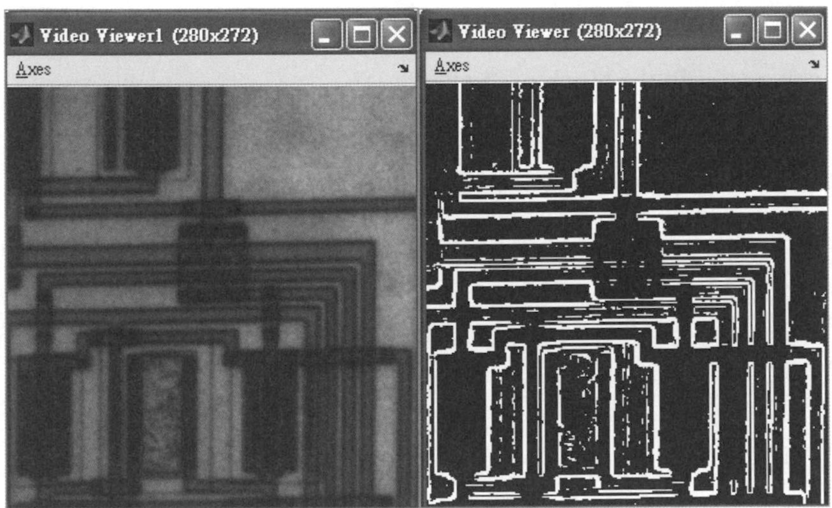

F. Translation

1. Keyboard
2. Prompt
3. Density
4. Liquid

5. Capacitor
6. Transform
7. Voltage
8. Direct
9. Enlarge
10. Extend

G. Questions

1. What will be changed in the output image, if we select the check box of "Edge Thinning" in the block of "Sobel" (2)?
2. What will be changed in the output image, if we enter "3" in the threshold scale factor in the block of "Sobel" (2)?
3. What will be changed in the output image, if we enter "10" in the threshold in the block of the "Find Local Maxima" (5)?

Image Analysis and Enhancement
圖像分析和強化

6-1 Image Sharpening
影像銳化

6-2 Cancellation of Salt and Pepper Noise
從影像中移除黑白點雜訊

6-3 Cancellation of Pseudo Image in Video Signal
移除影片中的虛擬影像

6-1 Image Sharpening

A. Command Window

```
I= im2double(imread('peppers.png'));

imshow(I)
```

B. SimuLink Layout

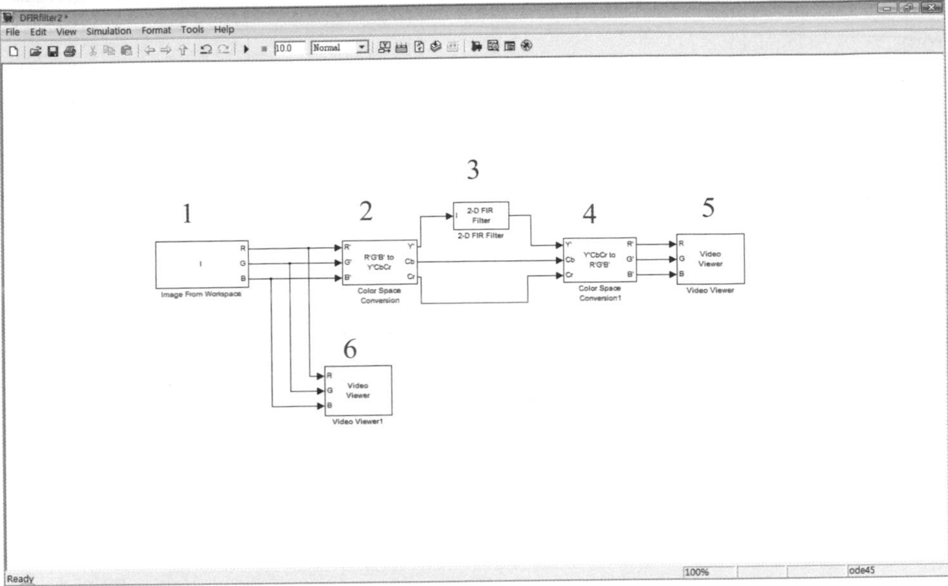

C. Path of Block

1. Video and Image Processing Blockset→Sources
   →Image From Workspace
2. Video and Image Processing Blockset→Conversions
   →Color Space Conversion
3. Video and Image Processing Blockset→Filtering→2-D FIR Filter
4. Video and Image Processing Blockset→Conversions
   →Color Space Conversion
5. Video and Image Processing Blockset→Sinks→Video Viewer

D. Parameter Settings

(1)Image From Workspace	(2)Color Space Conversion	(3)2-D FIR Filter	(4)Color Space Conversion	(5)Video Viewer	(6)Video Viewer
Main pane, Value = I	Conversion = R'G'B' to Y'CbCr	Coefficients = fspecial('unsharp')	Conversion = Y'CbCr to R'G'B'	Default	Default
Output port labels = R'|G'|B'		Output size = Same as input port I			
		Padding options = Symmetric			
		Filtering based on = Correlation			

E. Results

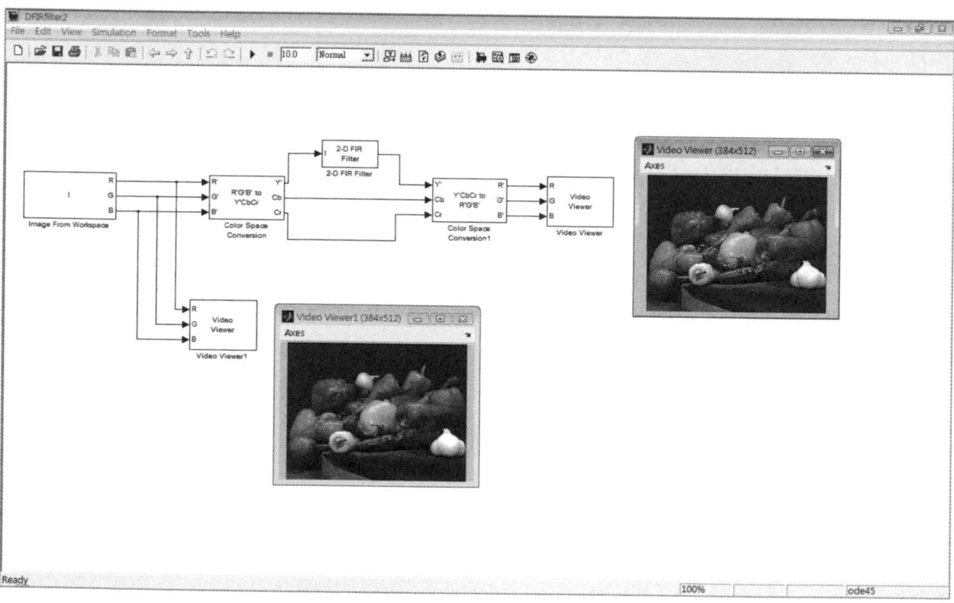

F. Translation

1. Conversion
2. Simulation
3. Toolbox
4. Sink
5. Filter
6. Preference
7. Coefficient
8. Control
9. Main
10. Sharp

G. Question

1. What will be changed in the output image, if we add a new “2-D FIR filter” between Cr signals in the blocks (2) and (4)?

6-2 Cancellation of Salt and Pepper Noise

A. Command Window

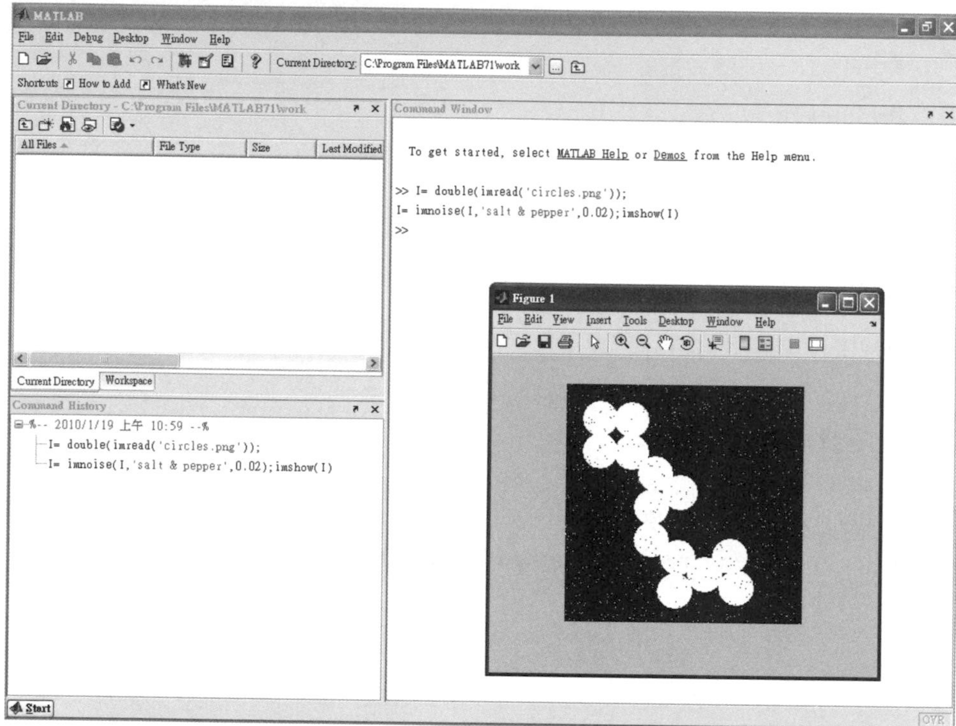


```
I= double(imread('circles.png'));

I= imnoise(I,'salt & pepper',0.02);

imshow(I)
```

B. SimuLink Layout

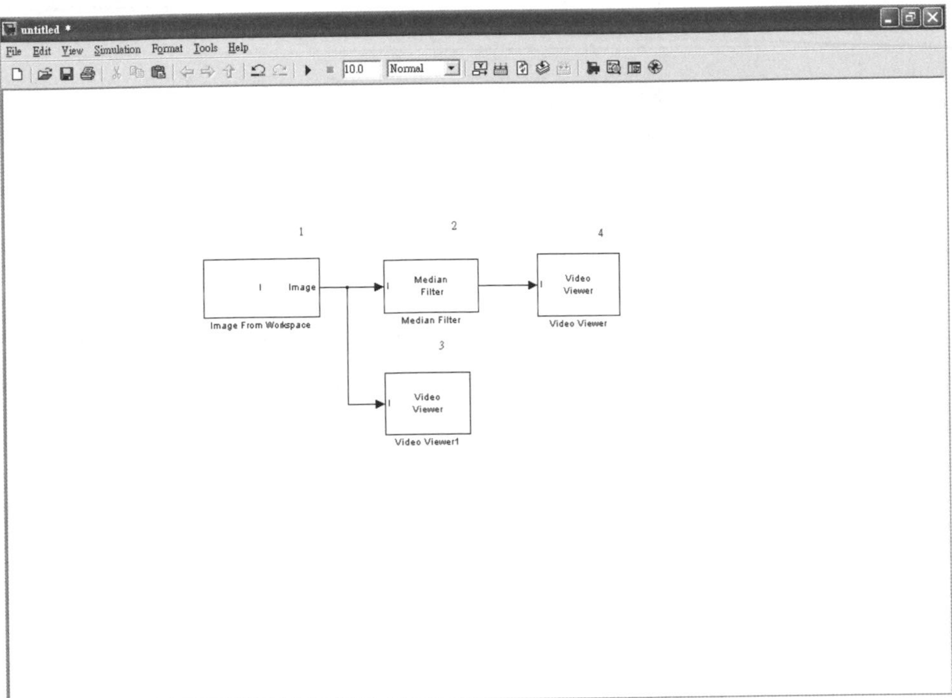

C. Path of Block

1. Video and Image Processing Blockset→Sources→Mage From Workspace
2. Video and Image Processing Blockset→Filtering→Median Filter
3. Video and Image Processing Blockset→Sinks→Video Viewer
4. Video and Image Processing Blockset→Sinks→Video Viewer

D. Parameter Settings

(1)Mage From Workspace	(2)Median Filter	(3)Video Viewer	(4)Video Viewer
Value = I	Default	Input image type= Intensity	Input image type= Intensity
Output port labels = Image			

E. Results

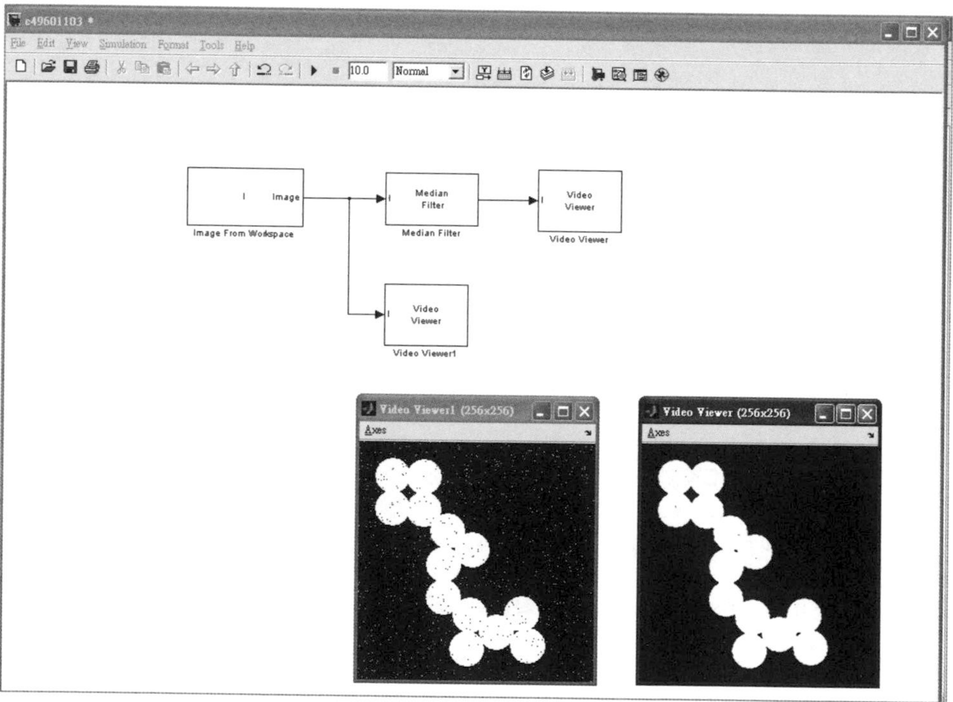

F. Translation

1. Individual
2. Noise
3. Pepper
4. In addition

5. Remove
6. Original
7. Noisy
8. Salt
9. Perfect
10. Assemble

G. Question

1. How can we make the output image perfect by adjusting the value of "Neighborhood size" in the "median filter" (2)?

6-3 Cancellation of Pseudo Image in Video Signal

A. SimuLink Layout

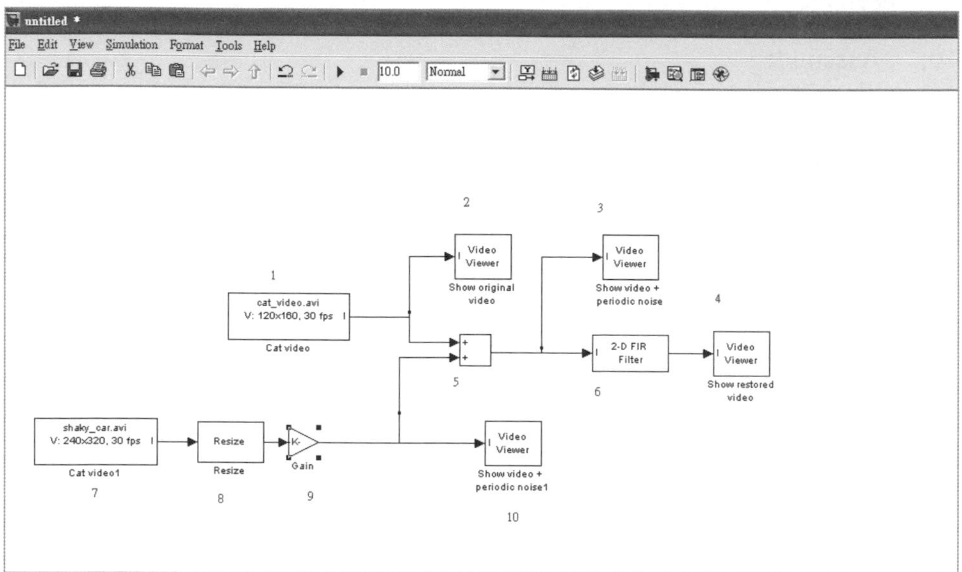

B. Path of Block

1. Video and Image Processing Blockset→Sources→Read AVI File
2. Video and Image Processing Blockset→Sinks→Video Viewer
3. Video and Image Processing Blockset→Sinks→Video Viewer
4. Video and Image Processing Blockset→Sinks→Video Viewer
5. Simulink→Math Operations→Add
6. Video and Image Processing Blockset→Filtering→2-D FIR Filter
7. Video and Image Processing Blockset→Sources→Read AVI File
8. Video and Image Processing Blockset→Geometric Transformations →Resize

9. Simulink→Commonly Used Blocks→Gain
10. Video and Image Processing Blockset→Sources→Read AVI File

C. Parameter Settings

(1)Cat video	(2)~(4)Show original video	(5)Add	(6)2-D Histogram
Input file name=cat_video.avi	Input image type=Intensity	default	Coefficient=0.8
Video Output data type=double			Output size=same as input port I Padding Cptions=Cicalar

(7)Cat video	(8)Reshape	(9)Gain	(10)Show original video
Input file name= shaky_car.avi	Numbers of output rows and columns=[120.160]	Gain=0.25	Input image type=Intensity
Video Output data type=double			

D. Results

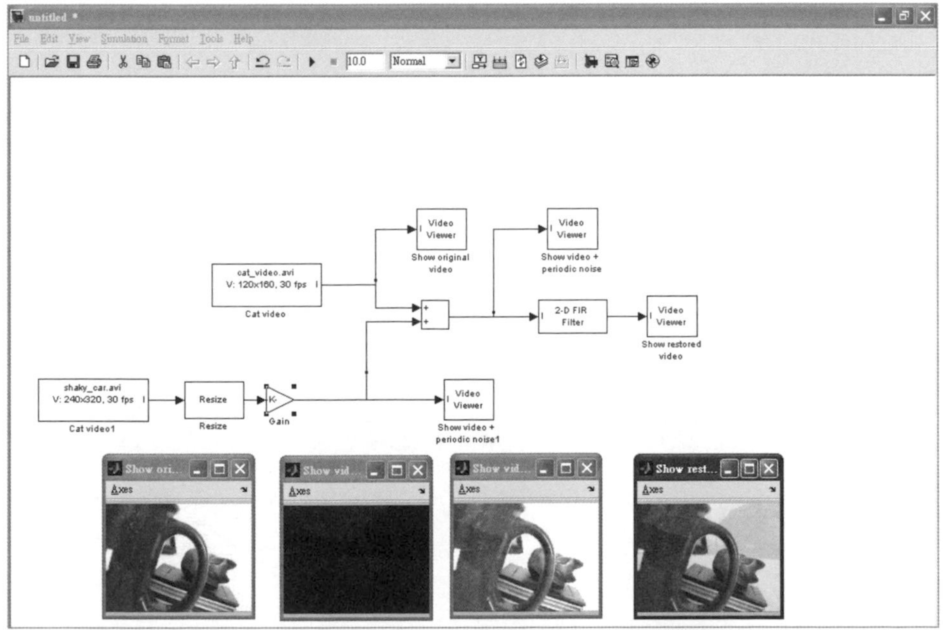

E. Translation

1. Target
2. Trace
3. Correlation
4. Information
5. Identification
6. Recognition
7. Presentation
8. Introduction
9. Realize
10. Framework

F. Question

1. Please adjust the value of gain in block (9) to make the output image perfect.

Pixel Statistics 像素統計

7-1 Image Histogram Computation

A. Command Window

```
I= im2double(imread('peppers.png'));
imshow(I)
```

B. SimuLink Layout

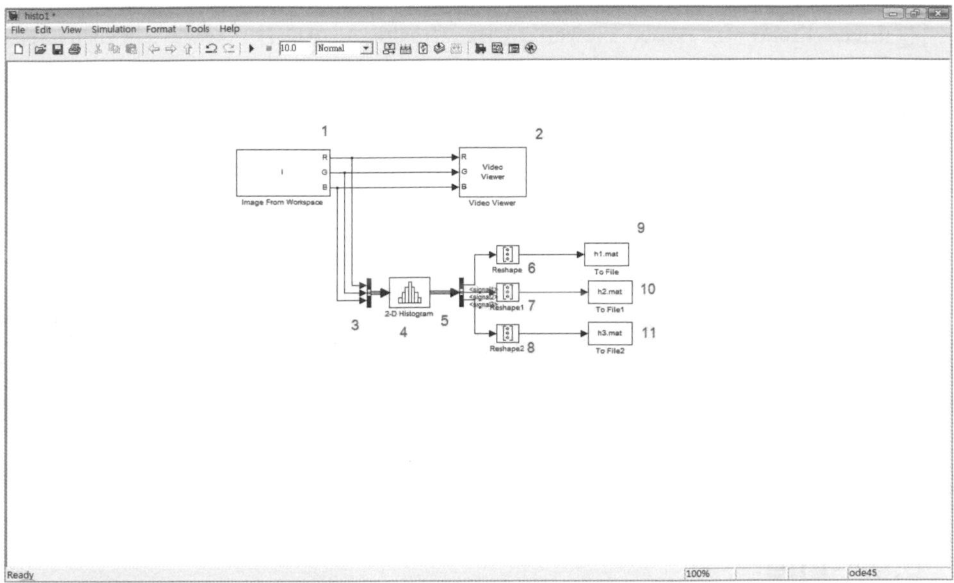

C. Part Locations

1. Video and Image Processing Blockset→Sources→Image From Workspace
2. Video and Image Processing Blockset→Sinks→Video Viewer
3. Simulink→Signal Routing→Bus Creator
4. Video and Image Processing Blockset→ Statistics→ 2-D Histogram
5. Simulink→Signal Routing→Bus Selector
6. Simulink →Math Operations→ Reshape
7. Simulink →Math Operations→ Reshape
8. Simulink →Math Operations→ Reshape
9. Simulink →Sinks→To File
10. Simulink →Sinks→To File
11. Simulink →Sinks→To File

D. Parameters settings

(1)Image From Workspace	(2)Video Viewer	(3)Bus Creator	(4)2-D Histogram	(5)Bus Selector
Value = I	Default	Number of inputs = 3	Default	Select signal 3

(6)~(8)Reshape	(9)To File	(10)To File	(11)To File
Output dimensionality =Column vector.	Filename=h1	Filename=h2	Filename=h3
	Variable name=h1	Variable name=h2	Variable name=h3

E. M-File

- File→New→m-file

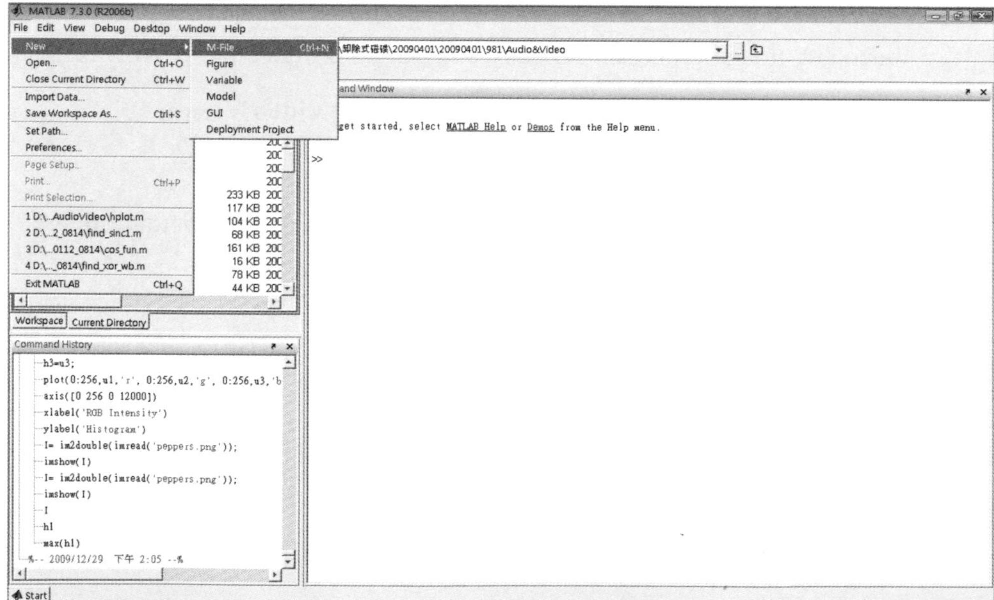

- Edit the code in m-file

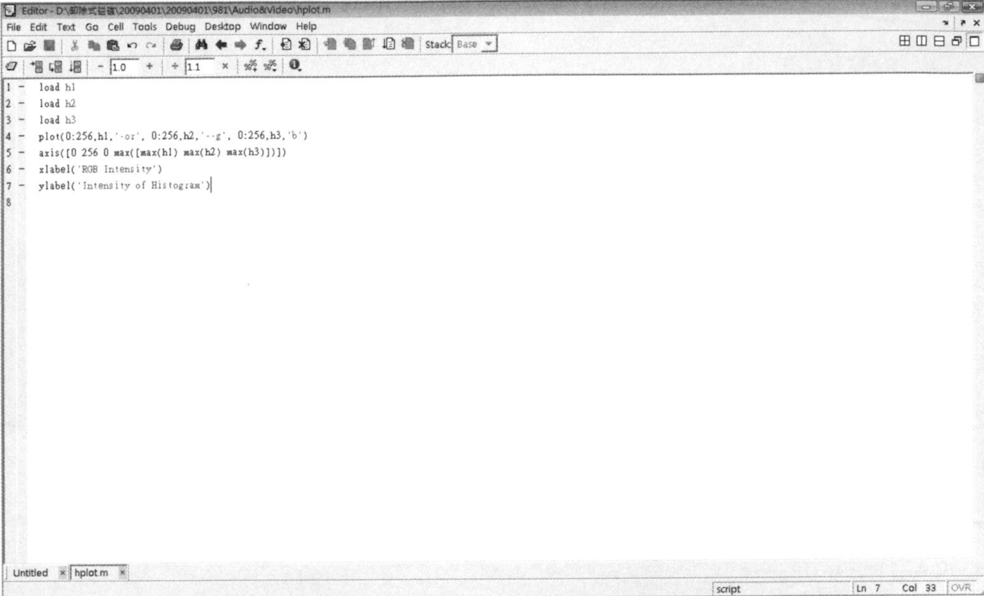

- % The h1(Red), h2(green), and h3(blue) will be plotted in the program
- % Save the program as the file name : hplot
- clear all;
- load h1
- load h2
- load h3
- plot(0:256,h1,'-or', 0:256,h2,'--g', 0:256,h3,'b')
- axis([0 256 0 max([max(h1) max(h2) max(h3)])])
- xlabel('RGB Intensity')
- ylabel('Intensity of Histogram')

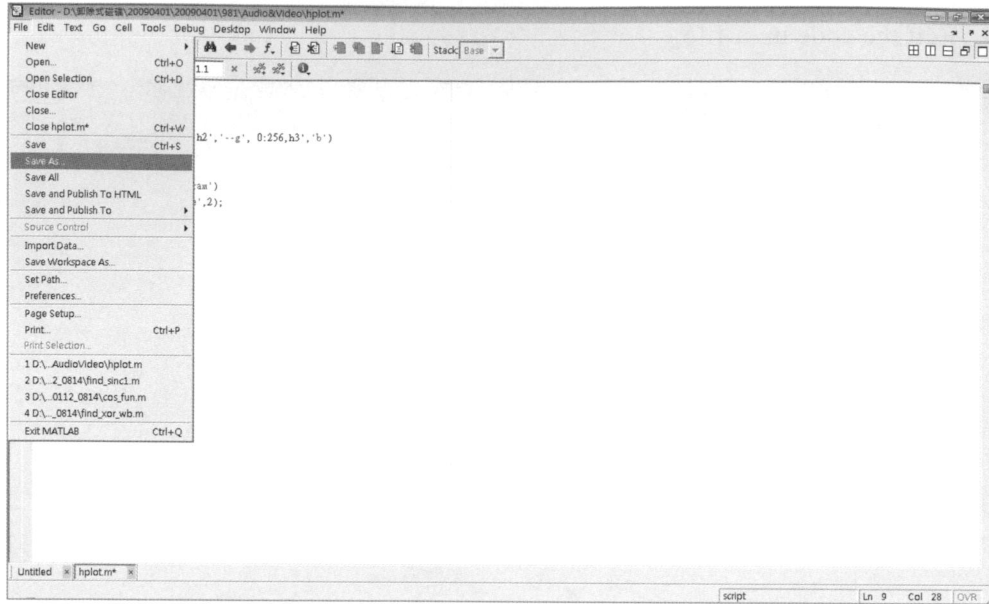

- Save as: hplot

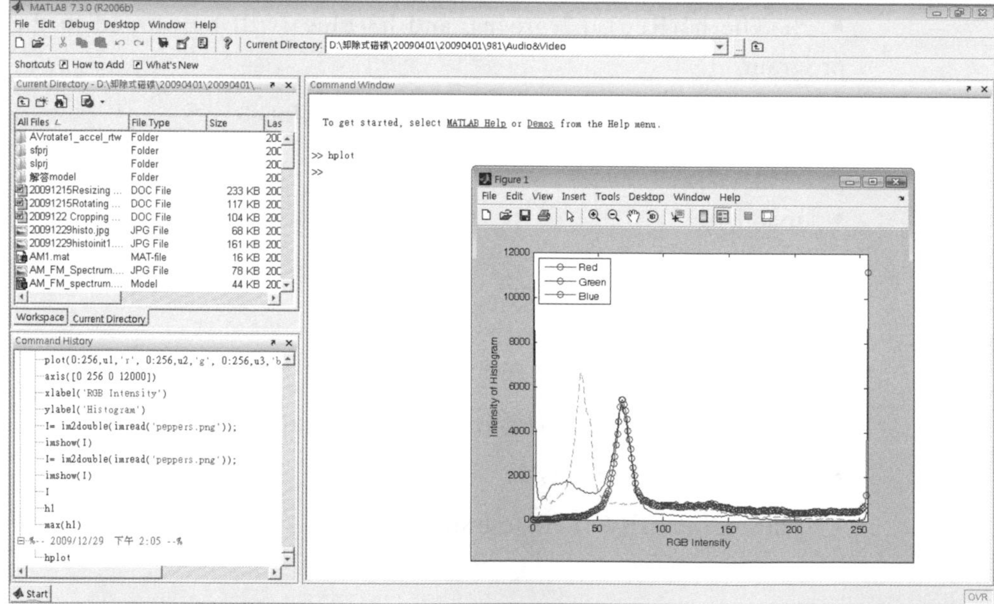

- Run the program: enter "hplot" in command window

F. Translation

1. Distribution
2. Statistics
3. Possibility
4. Histogram
5. Reshape
6. Multiplexer
7. Regular
8. Screen
9. Adjust
10. Pixel

G. Questions

1. What is "histogram"?
2. Download a picture, and compute its RGB histogram.

H. Reference

Examples of MatLab

NOTE 筆記頁

Pattern Matching 圖樣比對】

8-1 Tracking Object Using Correlation Computation

A. SimuLink Layout

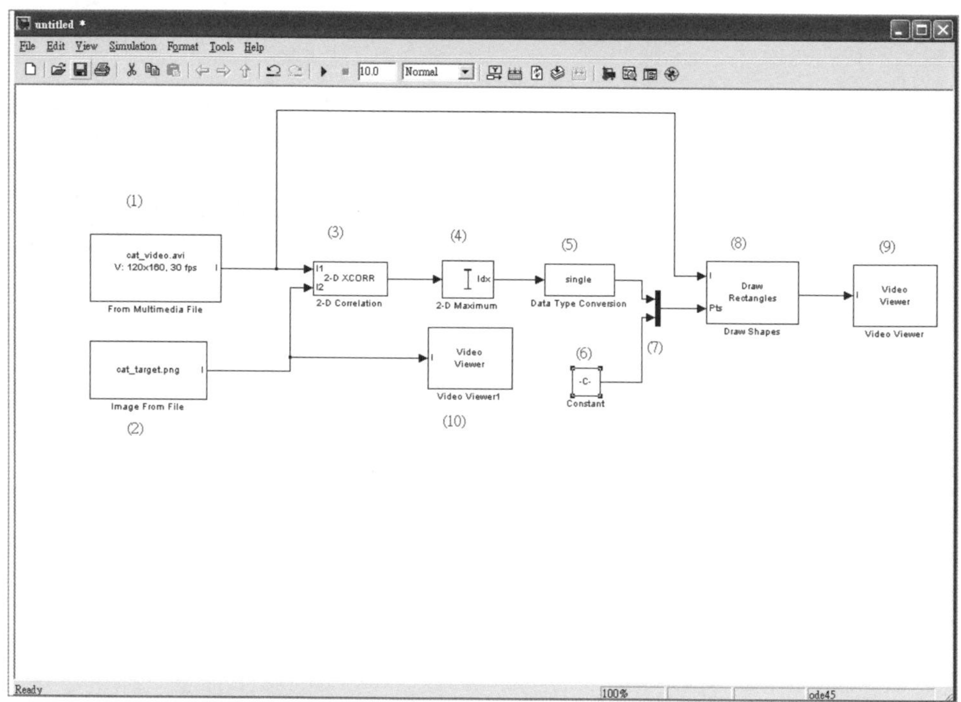

B. Path of Block

1. Video and Image Processing Blockset→Sources→From Multimedia File
2. Video and Image Processing Blockset→Sources→Image From File
3. Video and Image Processing Blockset→Statistics→2-D Correlation
4. Video and Image Processing Blockset→Statistics→2-D Maximum
5. Simulink→Signal Attributes→Data Type Conversion
6. Simulink→Sources→Constant

0 1 2 3 4 5 6 7 8 9 10

7. Simulink→Signal Routing→Mux
8. Video and Image Processing Blockset→Text & Graphics→Draw Shapes
9. Video and Image Processing Blockset→Sinks→Video Viewer
10. Video and Image Processing Blockset→Sinks→Video Viewer

C. Parameter Settings

(1)From Multimedia File	(2)Image From File	(3)2-D Correlation	(4)2-D Maximum	(5)Data Type Conversion
Input file name= cat_video.avi	File name = cat_target.png	Output size = Valid	Mode =Index	Output data type mode=single
Output intensity video =V	Output port labels = I	Normalized output =V		
	Output data type = single			

(6)Constant	(7)Mux	(8)Draw Shapes	(9)~(10)Video Viewer
Constant value =([41 41])	Default	Input image type = Intensity	Input image type =Intensity
Interpmet vector parameters as 1-D = x		Shape = Rectangles	

D. Results

E. Translation

1. From Multimedia File
2. Image From File
3. Independent
4. Peak
5. Data Type Conversion
6. Vibration
7. Touch
8. Panel
9. Computation
10. Signal Attributes

F. Question

1. If “Constant value” is entered to be “[20 20]” in Constant block (6), what will be find in the output?

Image Compression
圖像壓縮

9-1 Image Compression

A. Command Window

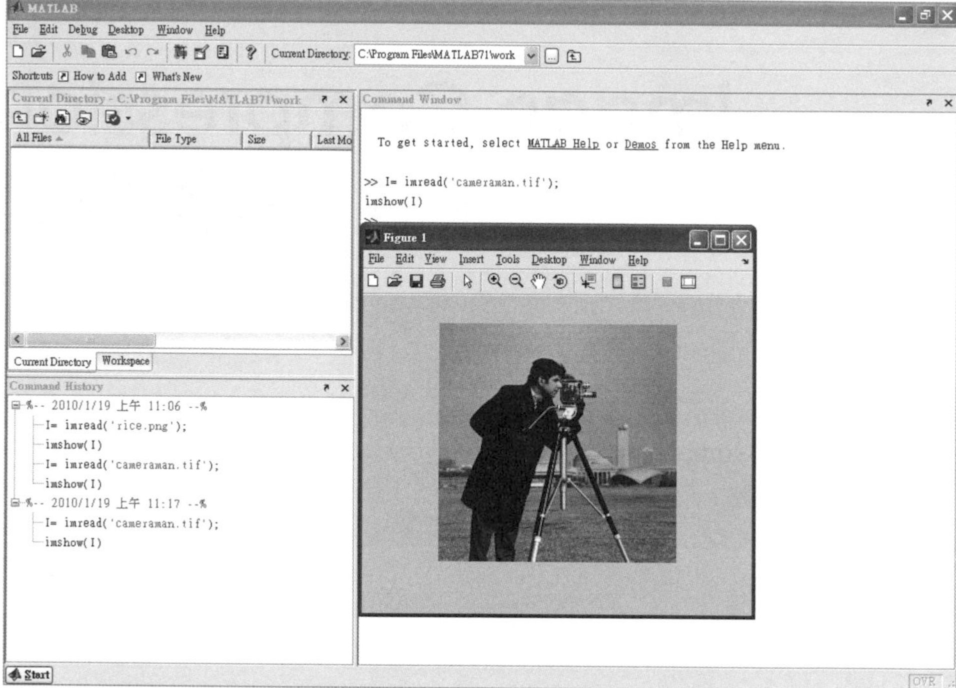


```
I= imread('cameraman.tif');

imshow(I)
```

B. SimuLink Layout

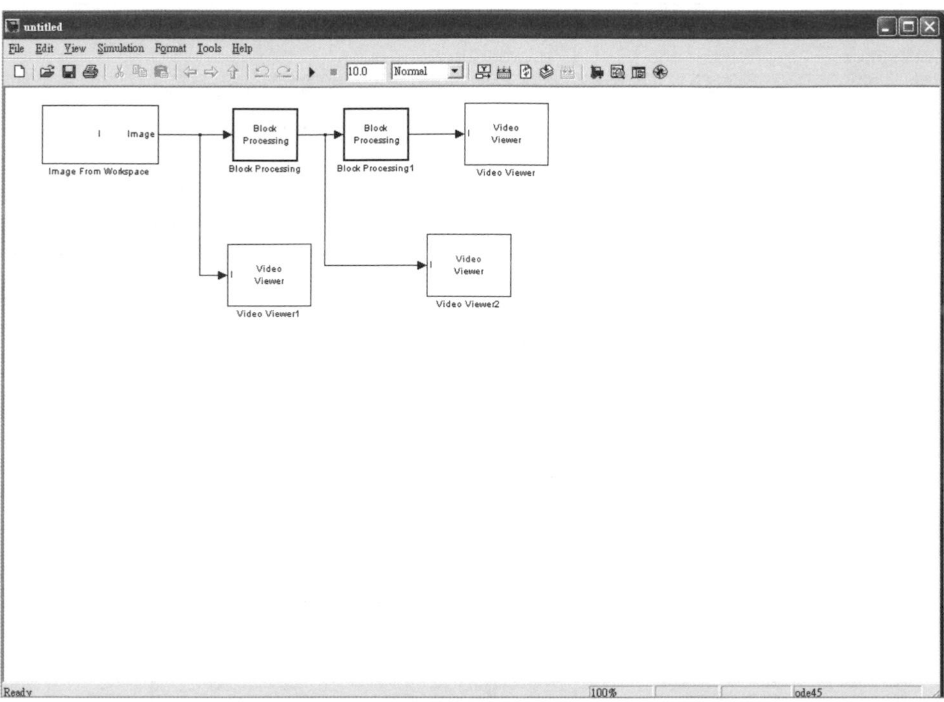

C. Path of Block

1. Video and Image Processing Blockset →SourcesVideo and Image → Image From Workspace
2. Video and Image Processing Blockset→UtilitiesVideo and Image →Block Processing
3. Video and Image Processing Blockset→UtilitiesVideo and Image →Block Processing
4. Video and Image Processing Blockset →Sinks→Video Viewer
5. Video and Image Processing Blockset →Sinks→Video Viewer

D. Parameter Settings

(1)Image From Workspace	(2)Block Processing	(3)Block Processing	(4)Video Viewer	(5)Video Viewer
Value = I Output port labels = Image Output data type = double	Open Subsystem	Default	Input image type=Intensity	Input image type=Intensity

E. SimuLink Layout in Subsystem

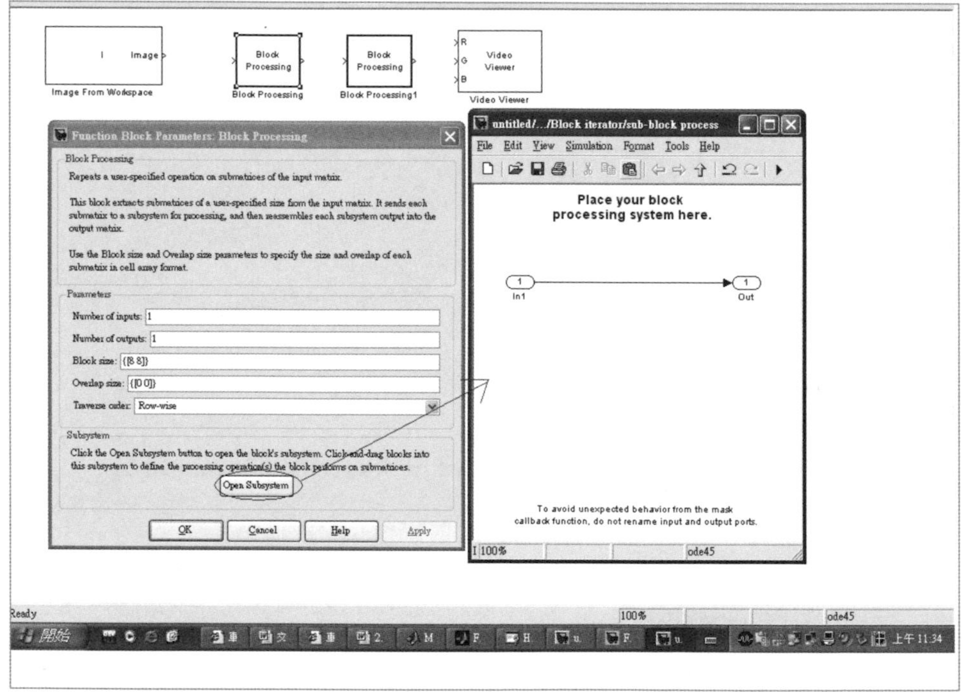

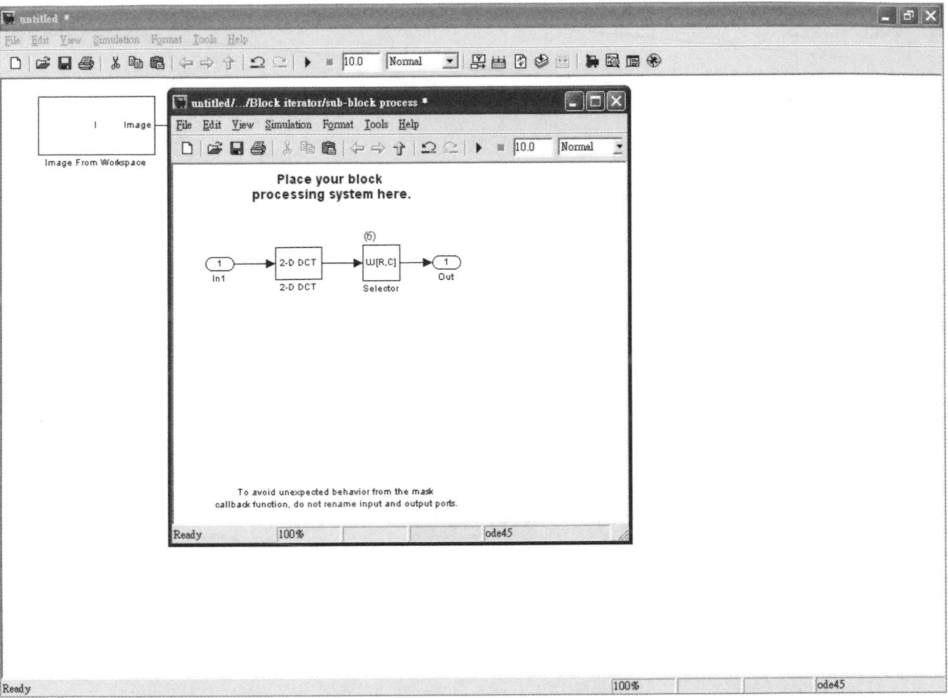

F. Part Locations

1. Video and Image Processing Blockset→Transforms→2-D DCT
2. Simulink→Signal Routing→Selector

(6)Selector
Input type = Matrix
Index mode = Zero-based
Rows (-1 for all rows) = 0
Columns (-1 for all columns) = 0
Select the Use index as starting value check box.
Output port dimensions = [4 4]

G. Results

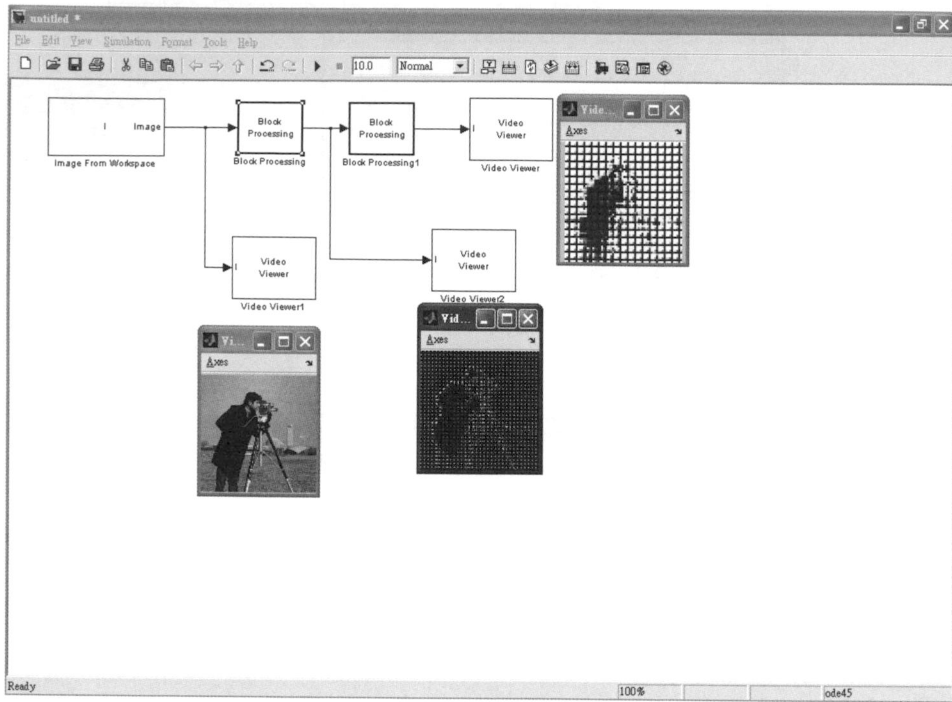

H. Translation

1. Double-click
2. Represent
3. Transmission
4. Submatrices
5. Subsystem
6. Before
7. Synthesis
8. Transfer
9. Bandwidth
10. Architecture

I. Questions

1. What will be changed in the output images if we set "overlap= {[4 4]}" in blocks (2) and (3)? Why?
2. How many pixels in the output images? Do you find the effect of image compression?

筆記頁

Modulation Technology
調變的技術

10-1 Amplitude / Frequency / Phase Modulation
振幅/頻率/相位調變

10-2 Amplitude/Frequency/Phase Modulation and Demodulation with Additive White Gaussian Noise (AM/FM/PM with AWGN)
加入式白高斯雜訊混入通道中的振幅/頻率/相位調變與解調

10-3 Comparison of AM and FM Spectrums
AM 與 FM 的頻譜之比較

10-4 Double-Sidedband AM and Double-Sidedband Suppressed-Carrier AM (DSB AM & DSBSC AM)
雙旁帶振幅調變與雙旁帶抑制載波振幅調變

10-1 Amplitude / Frequency / Phase Modulation

A. Simulink Layout

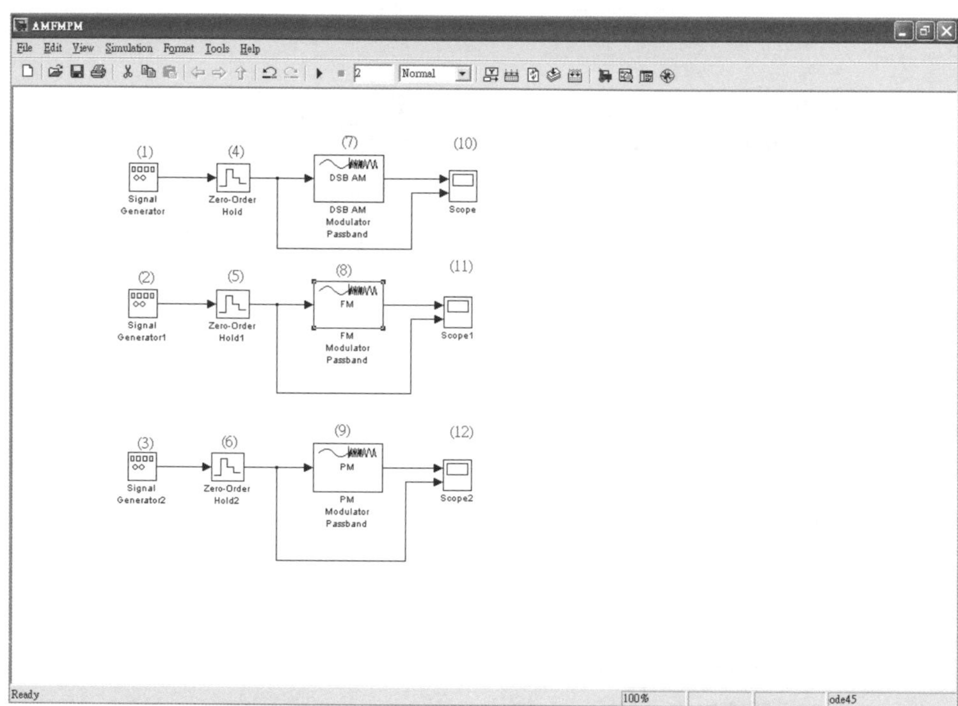

B. Path of the block

1. Simulink→Sources→Signal generator
2. Simulink→Sources→Signal generator
3. Simulink→Sources→Signal generator
4. Simulink→Discrete→Zero-order hold
5. Simulink→Discrete→Zero-order hold
6. Simulink→Discrete→Zero-order hold

7. Communication blockset→Modulation
→Analog passband modulation→ DSB AM Modulator Passband
8. Communication blockset→Modulation
→Analog passband modulation→ FM Modulator Passband
9. Communication blockset→Modulation
→Analog passband modulation→PM Modulator Passband
10. Simulink→Sinks→Scope
11. Simulink→Sinks→Scope
12. Simulink→Sinks→Scope

C. Parameter settings

(1)Signal generator	(2)Signal generator	(3)Signal generator	(4)Zero-order hold
Wave = sin	Wave = sin	Wave = square	Sample time = 1E-3

(5)Zero-order hold	(6)Zero-order hold	(7)DSB AM Modulator Passband	(8)FM Modulator Passband
Ssmple time = 0.001	Ssmple time = 0.001	Carrier frequency=10	Carrier frequency =10
			Frequency deviation=5

(9)PM Modulator Passband	(10)Scope	(11)Scope	(12)Scope
Carrier frequency=10	Parameters→ Number of channel=2	Parameters→ Number of channel=2	Parameters→ Number of channel=2

D. Results

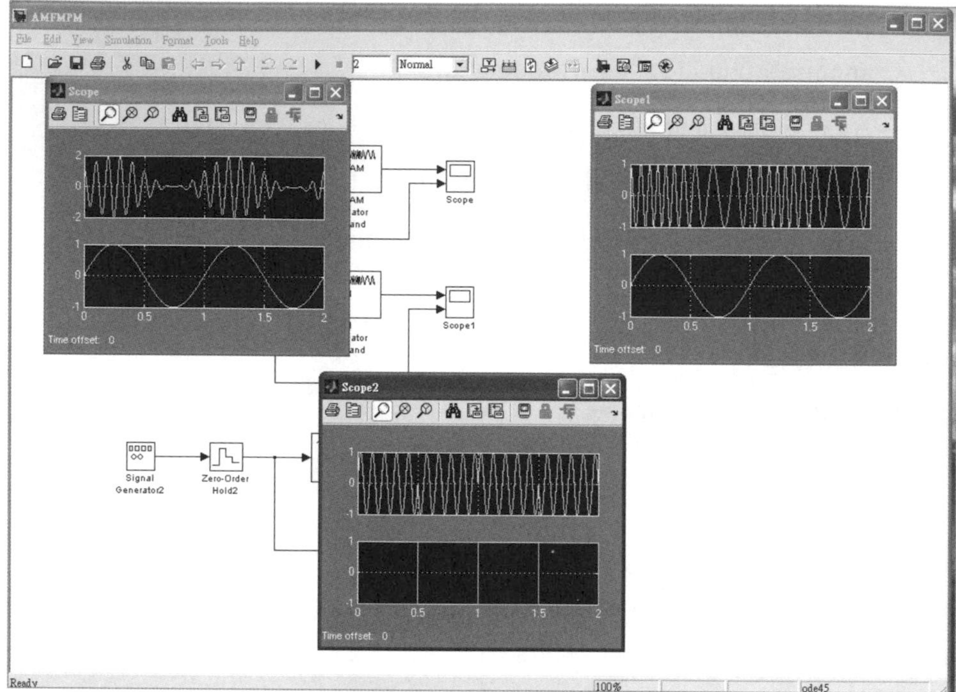

E. Translation

1. Amplitude
2. Frequency
3. Phase
4. Modulation
5. Deviation
6. Efficiency
7. Bandwidth
8. Demodulator
9. Scope
10. Carrier

F. Question

1. If the "Frequency deviation=3" in the block (8), what will be changed in the output signal of FM? Why?

10-2 Amplitude/Frequency/Phase Modulation and Demodulation with Additive White Gaussian Noise (AM/FM/PM with AWGN)

A. Simulink Layout

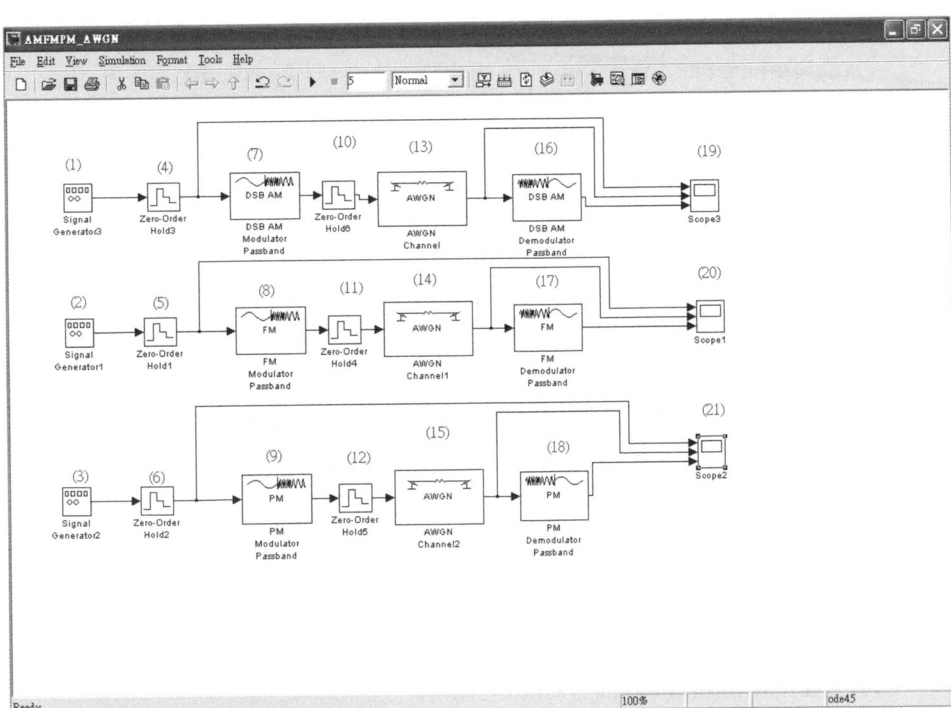

B. Path of the block

1. Simulink→Sources→Signal generator
2. Simulink→Sources→Signal generator
3. Simulink→Sources→Signal generator
4. Simulink→Discrete→Zero-order hold
5. Simulink→Discrete→Zero-order hold4
6. Simulink→Discrete→Zero-order hold
7. Communication blockset→Modulation
   →Analog passband modulation→DSB AM Modulator Passband
8. Communication blockset→Modulation
   →Analog passband modulation→FM Modulator Passband
9. Communication blockset→Modulation
   →Analog passband modulation→PM Modulator Passband
10. Simulink→Discrete→Zero-order hold
11. Simulink→Discrete→Zero-order hold
12. Simulink→Discrete→Zero-order hold
13. Communication blockset→Channel→AWGN Channel
14. Communication blockset→Channel→AWGN Channel
15. Communication blockset→Channel→AWGN Channel
16. Communication blockset→Modulation
    →Analog passband modulation→DSB AM Demodulator Passband
17. Communication blockset→Modulation
    →Analog passband modulation→FM Demodulator Passband
18. Communication blockset→Modulation
    →Analog passband modulation→PM Demodulator Passband

19. Simulink→Sinks→Scope
20. Simulink→Sinks→Scope
21. Simulink→Sinks→Scope

C. Parameter settings

(1)Signal generator	(2)Signal generator	(3)Signal generator	(4)Zero-order hold
Wave=sine	Wave=sine	Wave=square	Sampling time=0.001

(5)Zero-order hold	(6)Zero-order hold	(7)DSB AM Modulator Passband	(8)FM Modulator Passband
Sampling time=0.001	Sampling time=0.001	Carrier frequency=100	Carrier frequency=100
			Frequency deviation=60

(9)PM Modulator Passband	(10)Zero-order hold	(11)Zero-order hold	(12)Zero-order hold
Carrier frequency=100	Sampling time=1E-3	Sampling time=1E-3	Sampling time=1E-3

(13)AWGN Channel	(14)AWGN Channel	(15)AWGN Channel	(16)DSB AM Demodulator Passband
Mode=SNR	Mode=SNR	Mode=SNR	Carrier frequency =100
SNR=30	SNR=30	SNR=30	Low pass filter= Butterworth
Input signal power=0.25	Input signal power=0.25	Input signal power=0.25	Filter order=4
			Cutoff frequency=100

(17)FM Demodulator Passband	(18)PM Demodulator Passband	(19)Scope	(20)Scope
Carrier=100	Carrier=100	Parameters→ Number of channel=3	Parameters→ Number of channel=3
Frequency deviation=60			

(21)Scope			
Parameters→ Number of channel=3			

D. Results

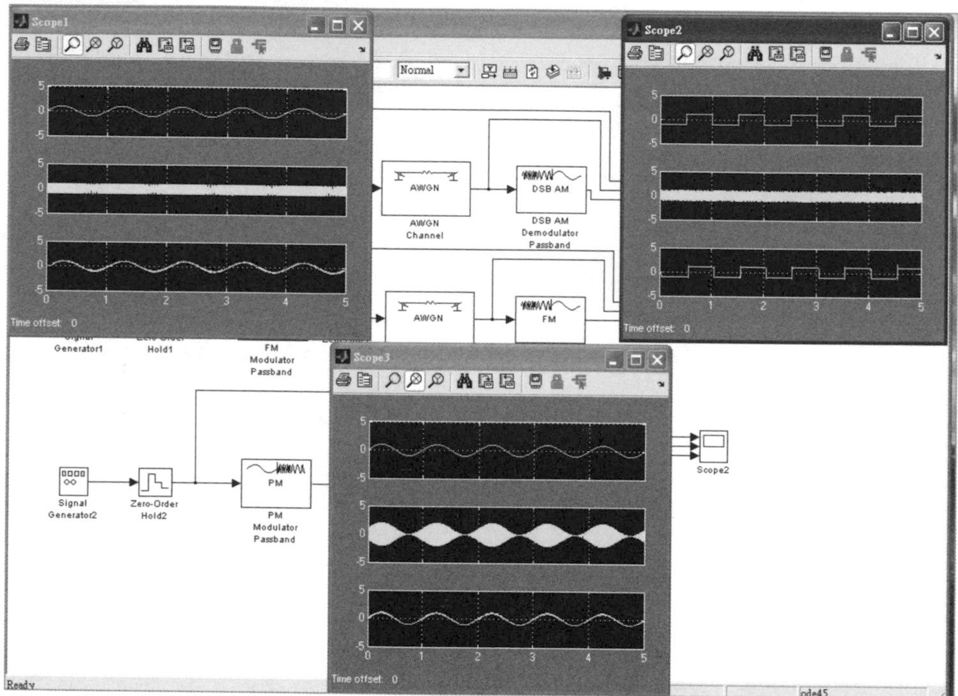

E. Translation

1. Additive
2. White
3. Gaussian
4. Noise
5. SNR: Noise to Signal Ratio
6. Channel
7. Hold

8. Discrete
9. Communication
10. Cutoff

F. Question

1. What is the output signals in blocks (19) to (21), if the "SNR=10" in the blocks (13) to (15)?

10-3 Comparison of AM and FM Spectrums

A. Simulink Layout:

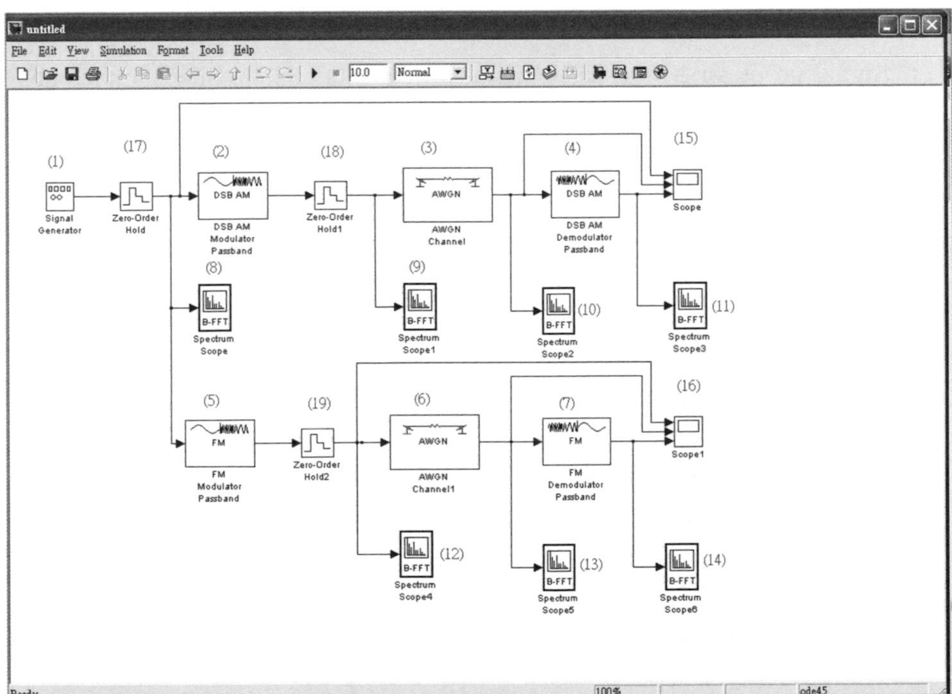

B. Path of the block

1. Simulink→Sources→Signal generator
2. Communication blockset→Modulation→Analog passband modulation→DSB AM Modulator Passband
3. Communication blockset→Channel→AWGN Channel
4. Communication blockset→Modulation→Analog passband modulation→DSB AM Demodulator Passband
5. Communication blockset→Modulation→Analog passband modulation→FM Modulator Passband
6. Communication blockset→Channel→AWGN Channel
7. Communication blockset→Modulation→Analog passband modulation→FM Demodulator Passband
8. Signal processing blockset→Signal processing sinks→Spectrum scope
9. Signal processing blockset→Signal processing sinks→Spectrum scope
10. Signal processing blockset→Signal processing sinks→Spectrum scope
11. Signal processing blockset→Signal processing sinks→Spectrum scope
12. Signal processing blockset→Signal processing sinks→Spectrum scope
13. Signal processing blockset→Signal processing sinks→Spectrum scope
14. Signal processing blockset→Signal processing sinks→Spectrum scope
15. Simulink→Sinks→Scope
16. Simulink→Sinks→Scope
17. Simulink→Discrete→Zero-order hold
18. Simulink→Discrete→Zero-order hold
19. Simulink→Discrete→Zero-order hold

C. Parameter settings

(1)Signal Generator	(2)DSB AM Modulator Passband	(3)AWGN Channel	(4)DSB AM Demodulator Passband	(5)FM Modulator Passband
Wave=sin	Carrier frequency =300	Mode=SNR	Carrier frequency =300	Carrier frequency=300
Amplitude=1		SNR=10	Low pass filter= Butterworth	frequency Deviation=60
Frequency =30		Input signal power=0.25	Filter order=4	
			Cutoff frequency =100	

(6)AWGN Channel	(7)FM Demodulator Passband	(8)~(14)Spectrum Scope	(15)~(16)Scope	(17)~(19)Zero-order Hold
Mode=SNR	Carrier frequency=300	Buffer input (v)	Parameters→ Number of channel=3	Sampling time=1E-3
SNR=10	frequency Deviation=60	Specify FFT Length (v)		
Input signal power=0.25	Filter order=200			

D. Results

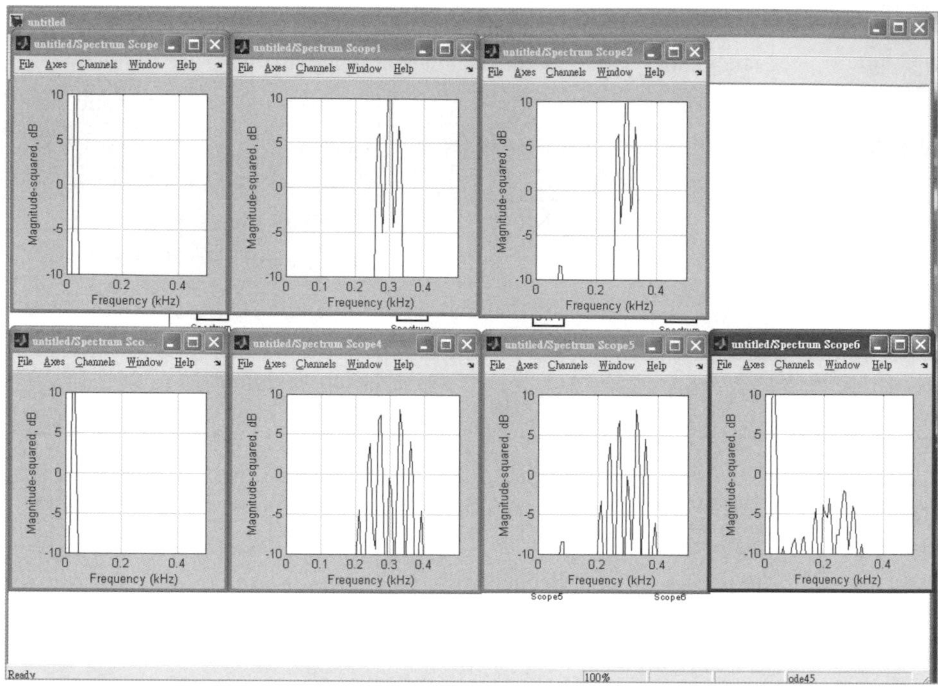
untitled
untitled/Spectrum Scope
untitled/Spectrum Scope1
untitled/Spectrum Scope2
untitled/Spectrum Sco...
untitled/Spectrum Scope4
untitled/Spectrum Scope5
untitled/Spectrum Scope6
File Axes Channels Window Help
Magnitude-squared, dB
Frequency (kHz)
Ready
100%
ode45

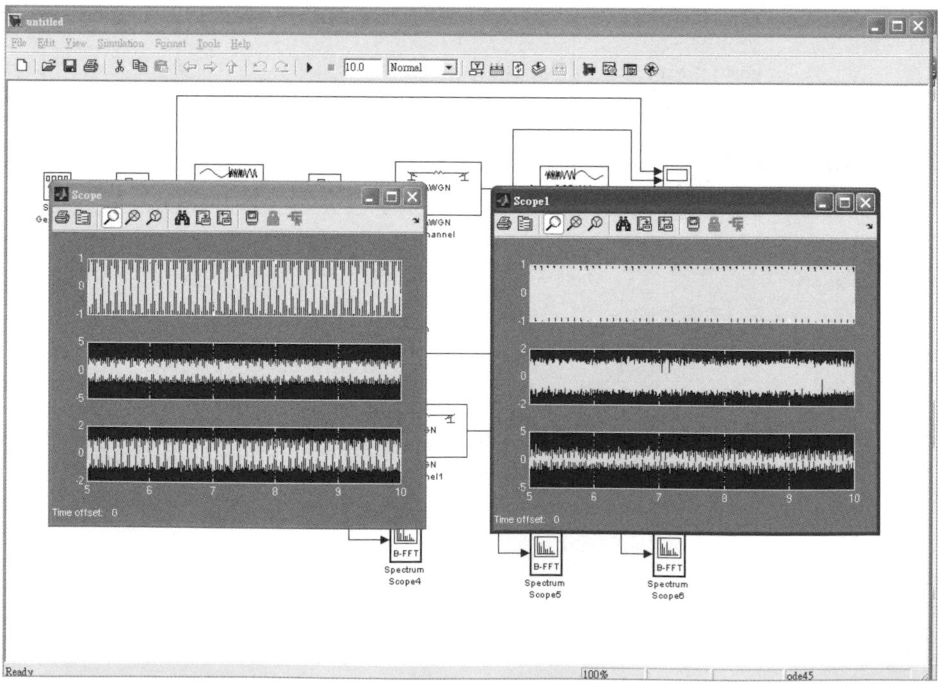
untitled
File Edit View Simulation Format Tools Help
10.0
Normal
Scope
Scope1
Time offset: 0
B-FFT
Spectrum Scope4
Spectrum Scope5
Spectrum Scope6
Ready
100%
ode45

E. Translation

1. Sideband
2. Index
3. Execution
4. Processor
5. Media
6. Analyzer
7. Spectrum
8. Interference
9. Signal processing
10. Simple

F. Questions

1. What are the differences of the spectrums of sine and square waves from Signal Generator?
2. Compare the results of SNR=10, and 30 in AWGN channels.
3. What are the differences of the filter order=4, and 10 in the results of AM Demodulations?
4. What are the differences of output signals of input signal power=0.25, and 0.05 from AWGN channels?
5. What are the differences of carrier frequencies=300 and 500 Hz in the results of AM, and FM demodulations?

*All the parameters select the first one of the two values, if the compared parameter was changed in each question.

10-4 Double-Sidedband AM and Double-Sidedband Suppressed-Carrier AM (DSB AM & DSBSC AM)

A. Simulink Layout

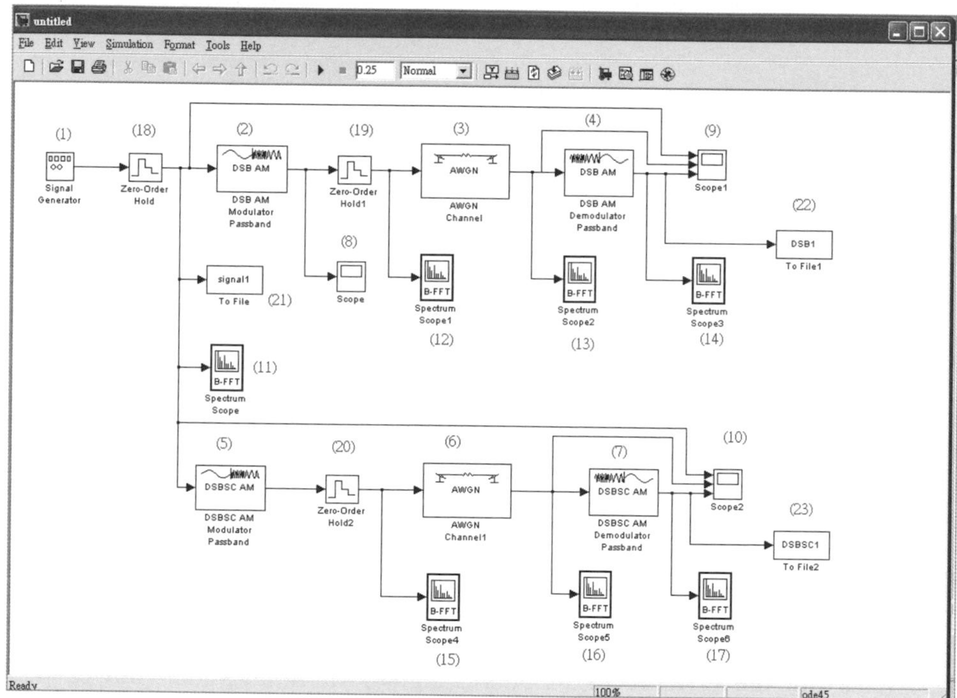

B. Path of the block

1. Simulink→Sources→Signal generator
2. Communication blockset→Modulation
   →Analog passband modulation→DSB AM Modulator Passband
3. Communication blockset→Channel→AWGN Channel

4. Communication blockset→Modulation
→Analog passband modulation→DSB AM Demodulator Passband
5. Communication blockset→Modulation
→Analog passband modulation→DSBSC AM Modulator Passband
6. Communication blockset→Channel→AWGN Channel
7. Communication blockset→Modulation
→Analog passband modulation→DSBSC AM Demodulator Passband
8. Simulink→Sinks→Scope
9. Simulink→Sinks→Scope
10. Simulink→Sinks→Scope
11. Signal processing blockset→Signal processing sinks→Spectrum scope
12. Signal processing blockset→Signal processing sinks→Spectrum scope
13. Signal processing blockset→Signal processing sinks→Spectrum scope
14. Signal processing blockset→Signal processing sinks→Spectrum scope
15. Signal processing blockset→Signal processing sinks→Spectrum scope
16. Signal processing blockset→Signal processing sinks→Spectrum scope
17. Signal processing blockset→Signal processing sinks→Spectrum scope
18. Simulink→Discrete→Zero-order hold
19. Simulink→Discrete→Zero-order hold
20. Simulink→Discrete→Zero-order hold
21. Simulink→Sinks→To file
22. Simulink→Sinks→To file
23. Simulink→Sinks→To file

C. Parameter settings

(1)Signal Generator	(2)DSB AM Modulator Passband	(3)AWGN Channel	(4)DSB AM Demodulator Passband
Wave=sine	Carrier frequency=300	Mode=SNR	Carrier frequency =300
Amp=0.2		SNR=10	Low pass filter= Butterworth
Freq=30		Input signal power=0.01	Filter order=4
			Cutoff frequency=100

(5)DSBSC AM Modulator Passband	(6)AWGN Channel	(7)DSBSC AM Demodulator Passband	(8)Scope
Carrier frequency =300	Mode=SNR	Carrier=300	Default
	SNR=10	Low pass filter= Butterworth	
	Input signal power=0.01	Filter order=4	
		Cutoff frequency =100	

(9)~(10)Scope	(11)~(17)Spectrum Scope	(18)~(20)Zero-order Hold	(21)To file
Parameters→ Number of channel=3	Buffer input (v)	Sampling time=1E-3	File name= signal1
	Specify FFT Length (v)		Variable name=s1

(22)To file	(23)To file		
File name= DSB1	File name= DSBSC1		
Variable name=DSB1	Variable name=DSBSC1		

D. Results

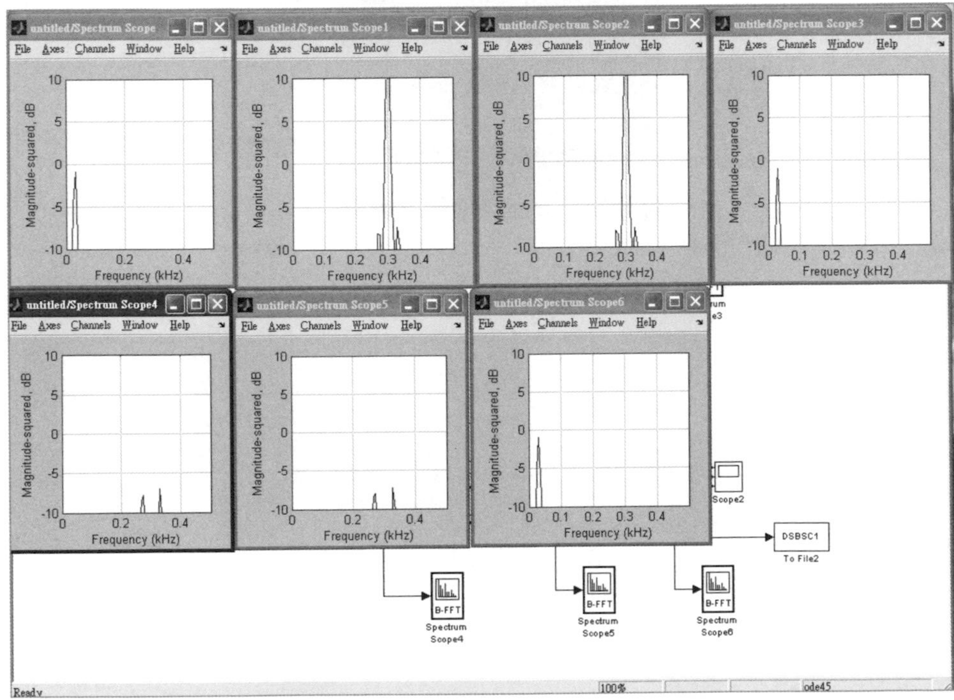

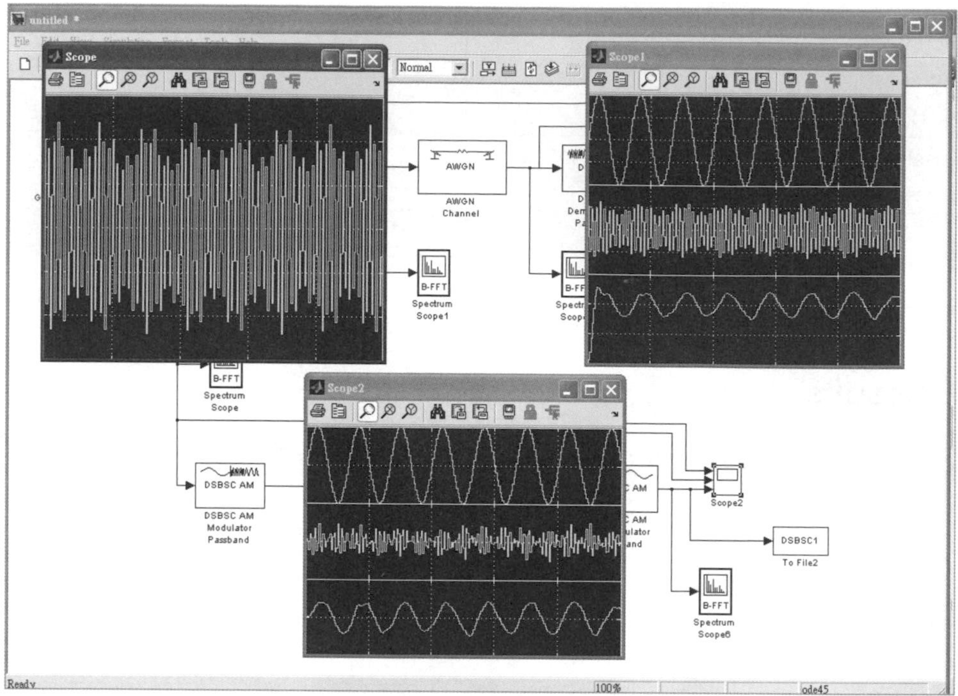

E. Translation

1. Sawtooth wave
2. Square wave
3. Sinusoidal wave
4. Analog
5. Digital
6. Order
7. DSB AM: Double-Sideband Amplitude Modulation
8. DSBSC AM: Double-Sideband Suppressed-Carrier Amplitude Modulation
9. Success
10. Continuous

F. Questions

*All the parameters select the first one of the two values, if the compared parameter was changed in each question.

1. What are the differences of the spectrums of sine, square, and sawtooth waves from Signal Generator?
2. Compute the modulation index (m) form scope (14). Amplitude= 0.2, 0.5, and 1 in signal generator (1) in the DSB AM system. (The other parameters were selected the first ones.)

   where
3. Find the main differences of DSB AM and DSBSC AM.
4. What do you find in the 2 figures from the results of the following program?

NOTE

筆記頁

Solution
問題解答

1-1 Import and View the AVI Files(Solution)
【1-1 輸入並觀看 AVI 檔案(問題解答)】

1. Keep running, and stop at the 120th second.
   【1.將程式改成持續執行 120 秒。】

- Hints(提示)

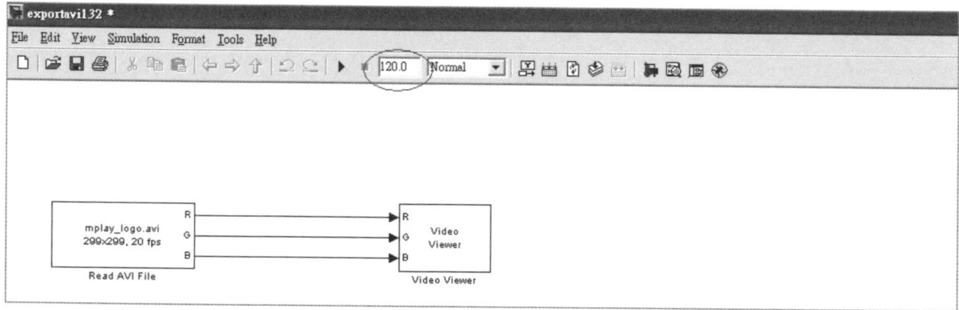

- Results(結果)

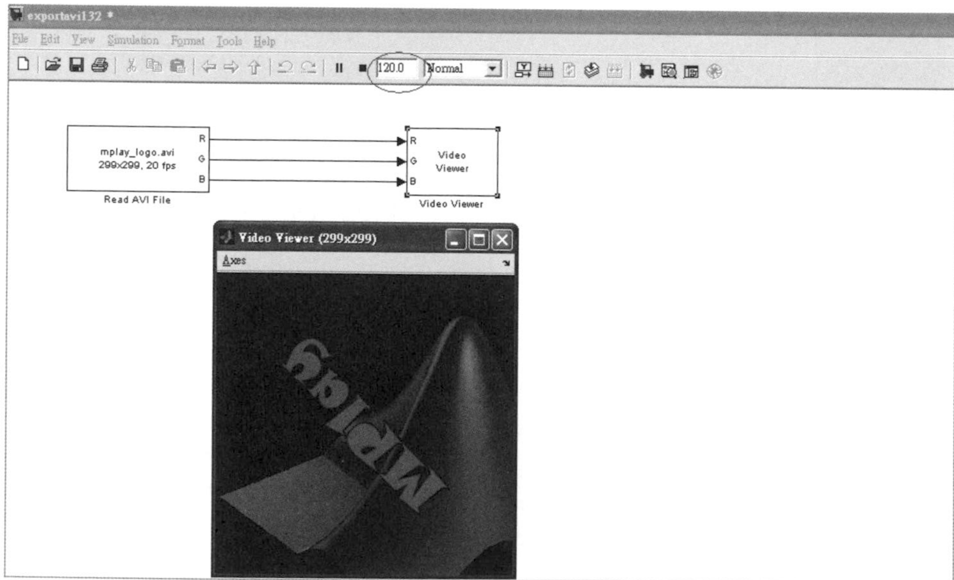

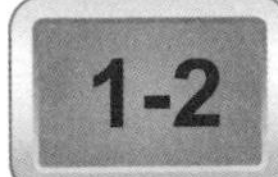

1-2 Export the AVI Files (Solution) 【1-2 輸出 AVI 檔案(問題解答)】

1. What is the output, if the "Gain=0.5" in Gain block (2)? Why?
   【1.如果改變增益方塊(2)裡面的"Gain(增益)= 0.5"，則輸出影像會有什麼改變？為什麼？】

- Hints(提示)

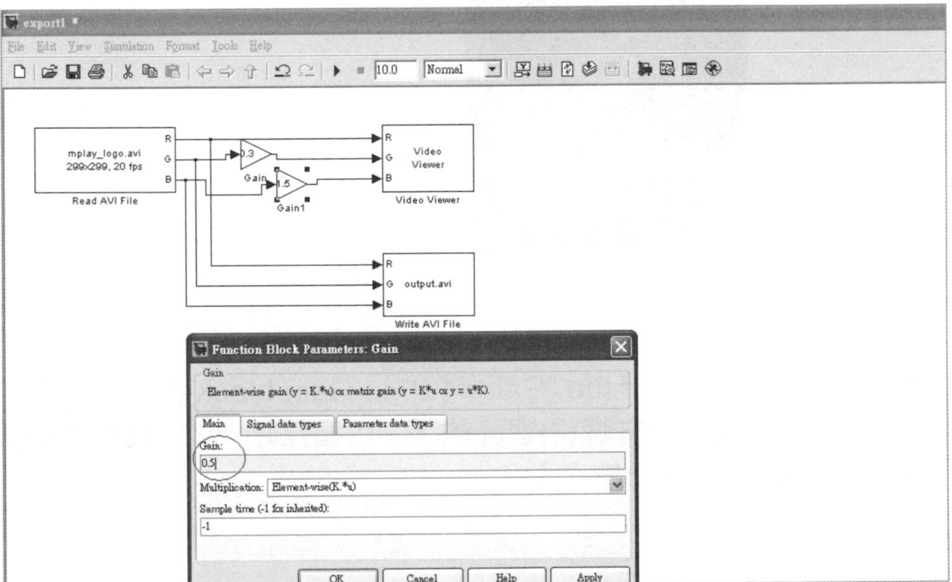

- Results(結果)

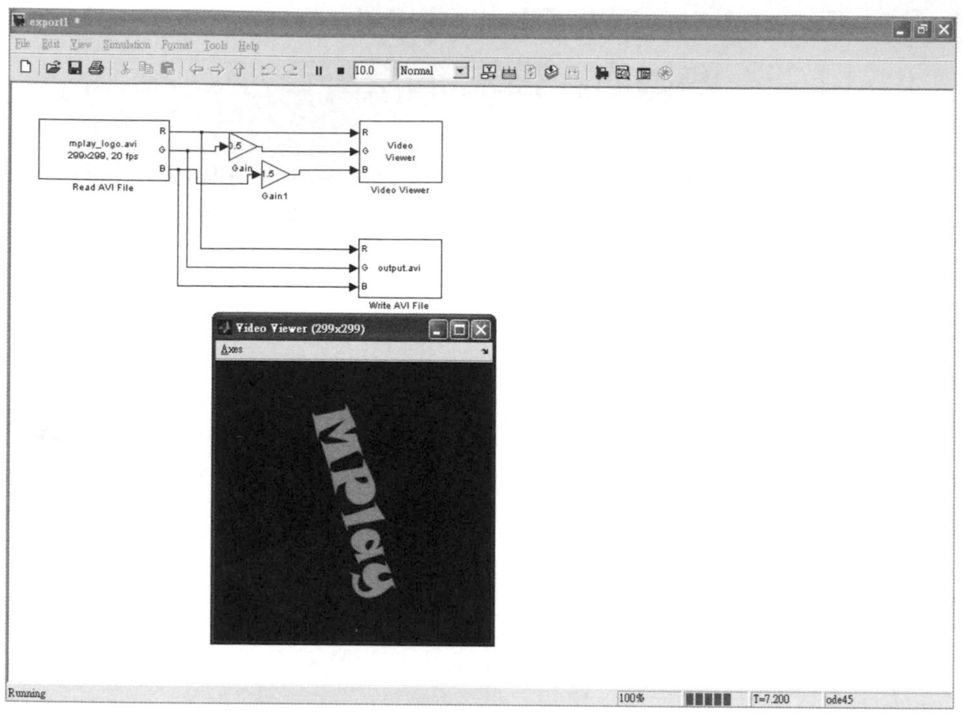

2. What is the output, if the “Gain=1.2” in Gain block (3)? Why?
   【2. 如果改變增益方塊(2)裡面的“Gain(增益)= 1.2”，則輸出影像會有什麼改變？為什麼？】

- Hints(提示)

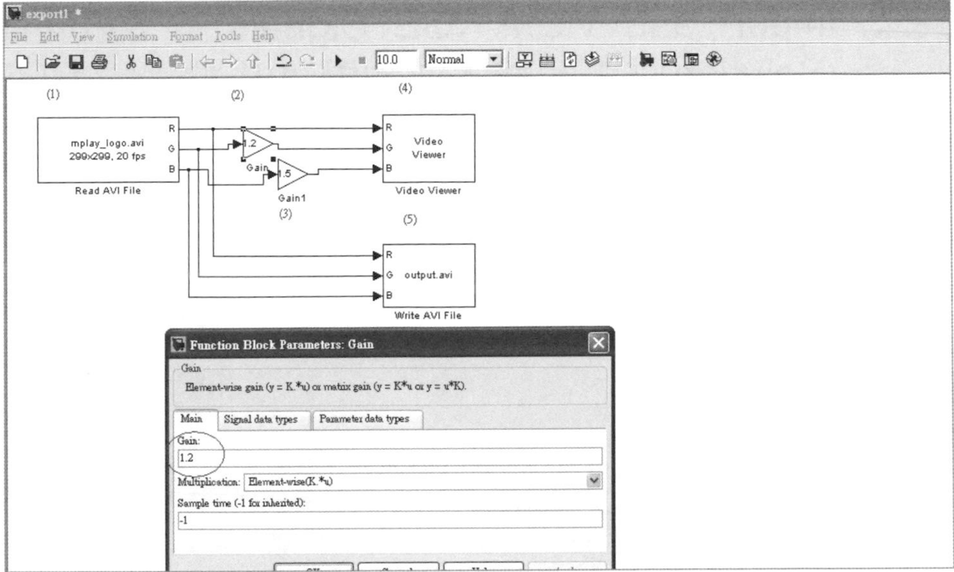

- Results(結果)

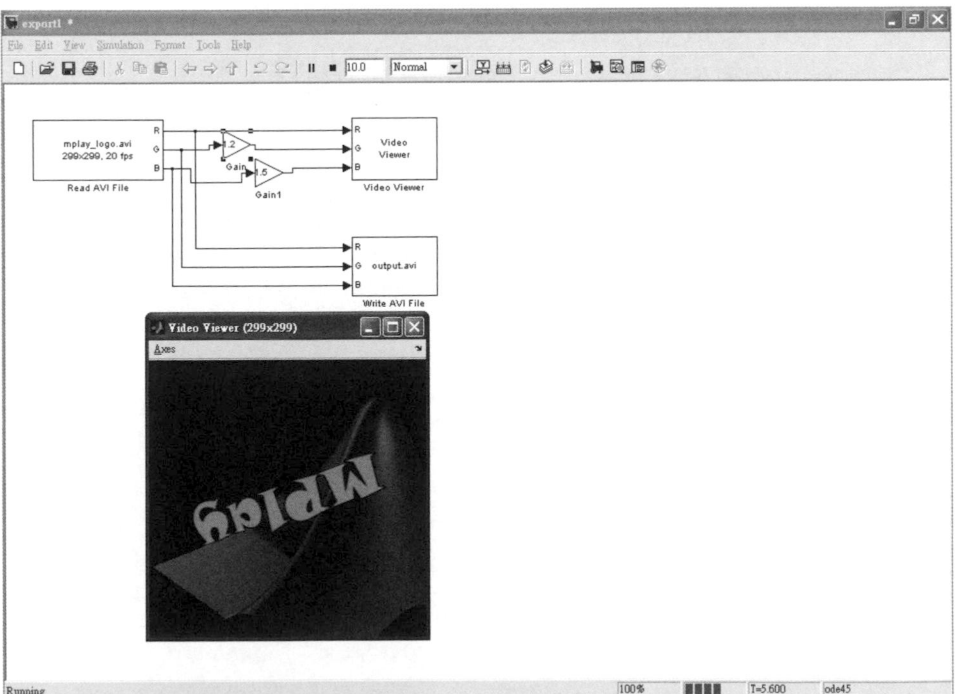

1-3 Import and View RGB Signals in AVI File (Solution)【1-3 輸入並觀看 AVI 檔案的 RGB 訊號(問題解答)】

1. If the links is changed to connect as follows:

   (a)R→R　G→G

   (b)R→R

   (c)R→G　G→B　B→R

   What will be displayed in the Video Viewer (2)?

【1. 當連接更改為以下方式：

   (a)R 接 R　G 接 G

   (b)R 接 R

   (c)R 接 G　G 接 B　B 接 R

   則在輸出的視頻播放器（2）將發生有何改變？】

- Hints(提示)

  (a) R→R

- Results(結果)

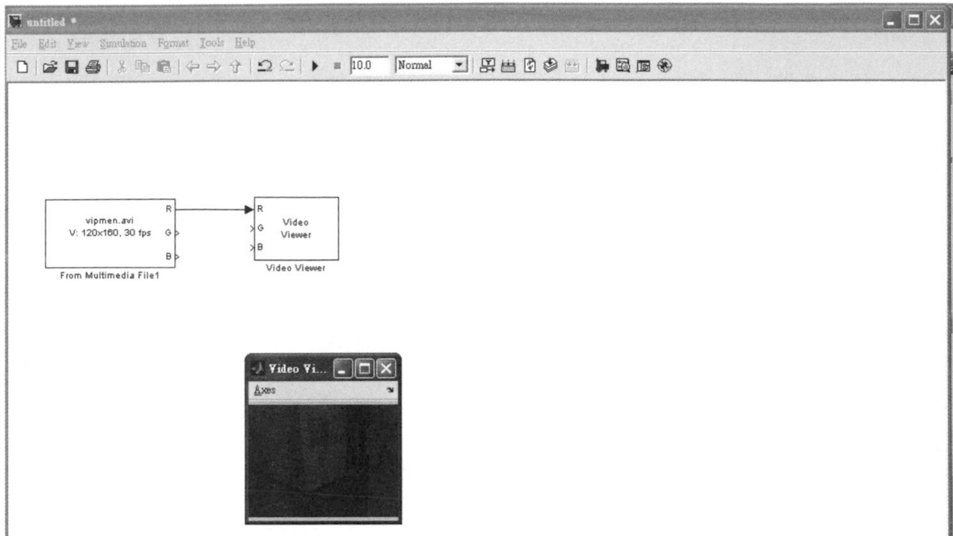

- Hints(提示)

  G→G

- Results(結果)

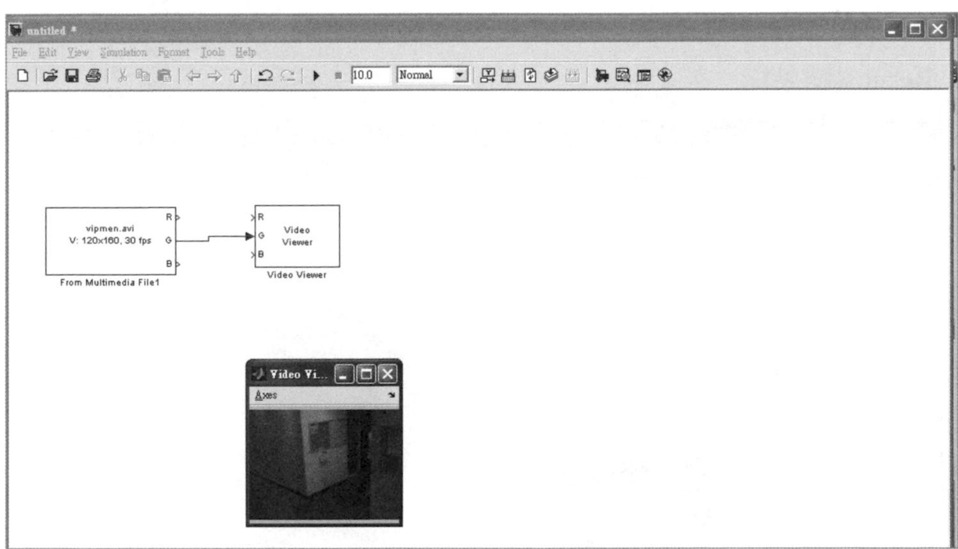

- Hints(提示)

 (b) R→R

- Results(結果)

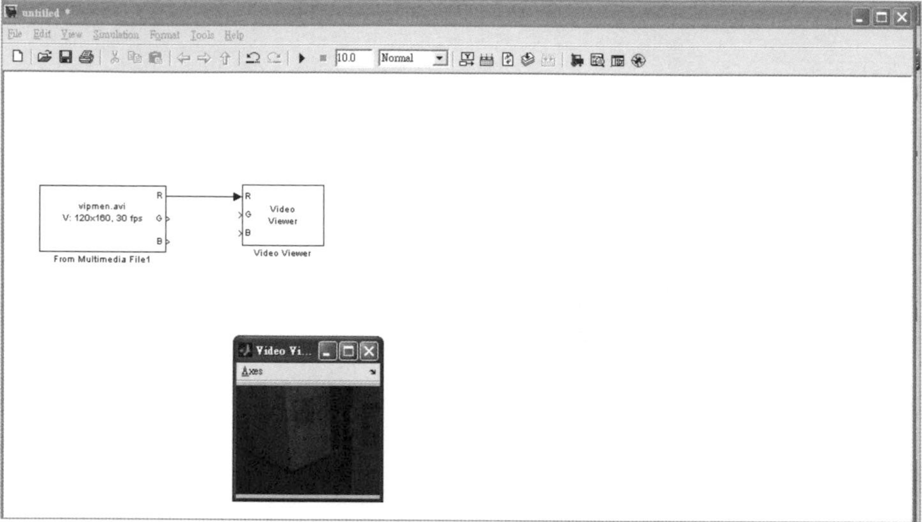

- Hints(提示)

 (c) R→G

- Results(結果)

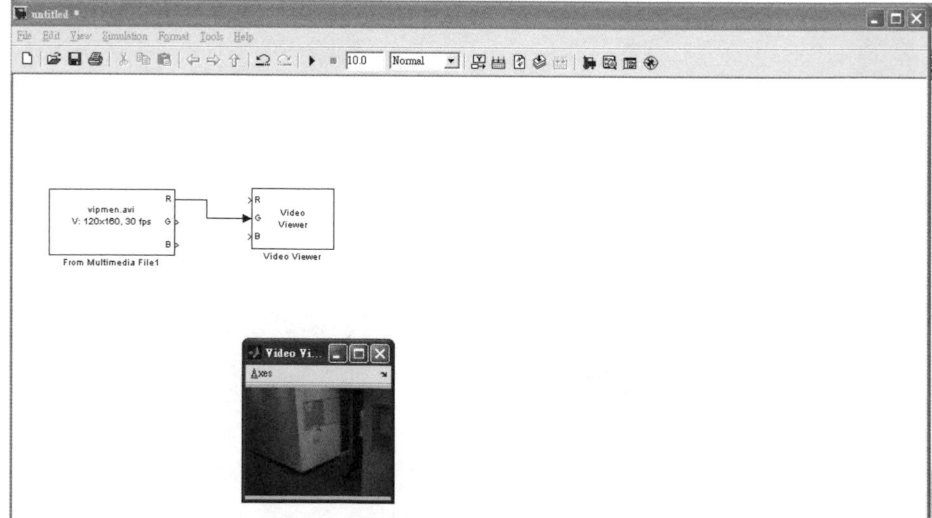

- Hints(提示)

  G→B

- Results(結果)

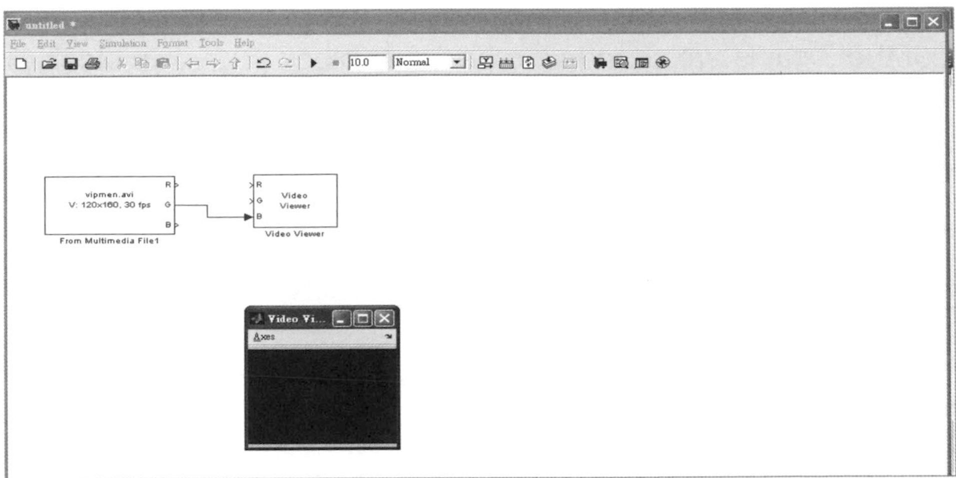

- Hints(提示)

  B→R

- Results(結果)

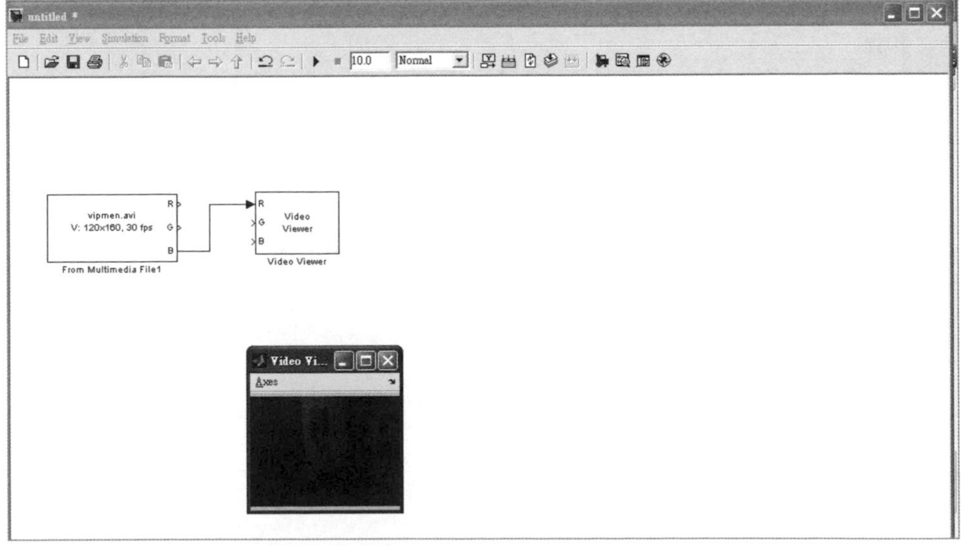

1-4 Export Multimedia Files (Solution)
【1-4 輸出到多媒體檔案(問題解答)】

1. If the 3 gains are changed to be 1.5 in block (2), 1.2 in block (3), and 1.5 in block (4), what will be the output?
   【1.在增益方塊(2)、(3)、(4)中，更改 "Gain(增益)=1.5(3)、1.2(4)、1.5(5)"，則輸出會有何變化？】

- Hints(提示)

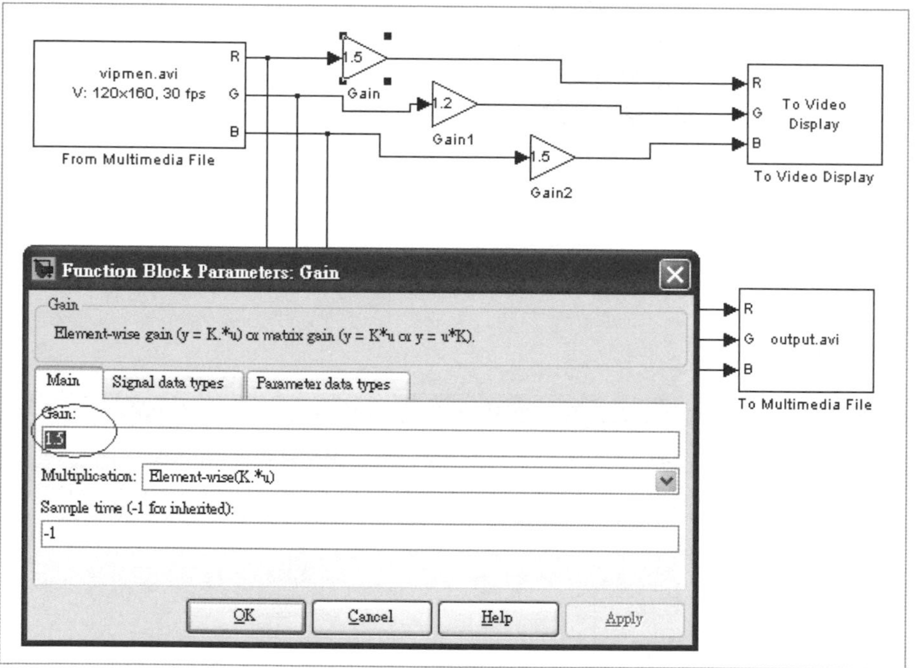

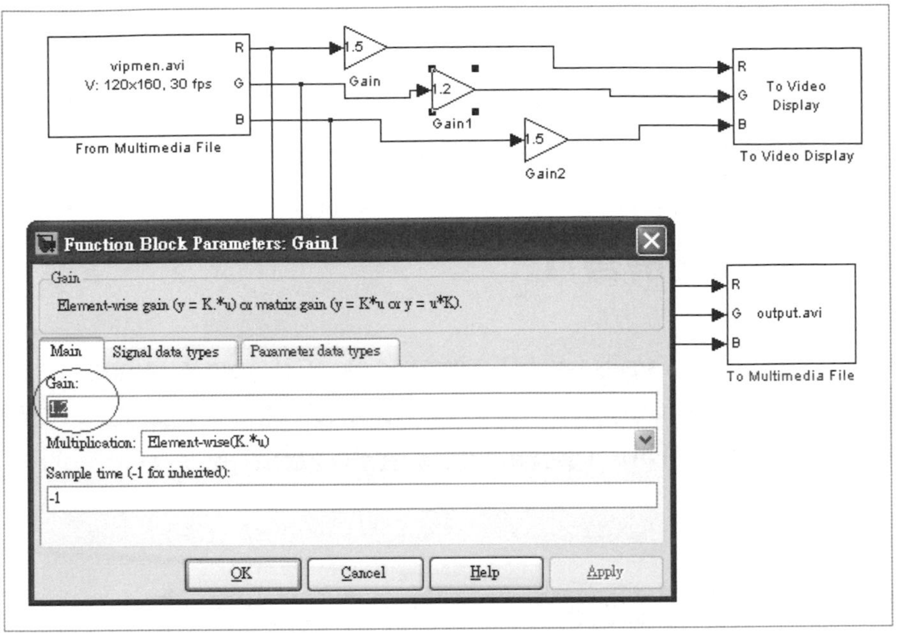
vipmen.avi
V: 120x160, 30 fps
R
G
B
From Multimedia File
1.5
Gain
1.2
Gain1
1.5
Gain2
To Video Display
To Video Display
Function Block Parameters: Gain1
Gain
Element-wise gain (y = K.*u) or matrix gain (y = K*u or y = u*K).
Main
Signal data types
Parameter data types
Gain:
1.2
Multiplication: Element-wise(K.*u)
Sample time (-1 for inherited):
-1
OK
Cancel
Help
Apply
output.avi
To Multimedia File

- Results(結果)

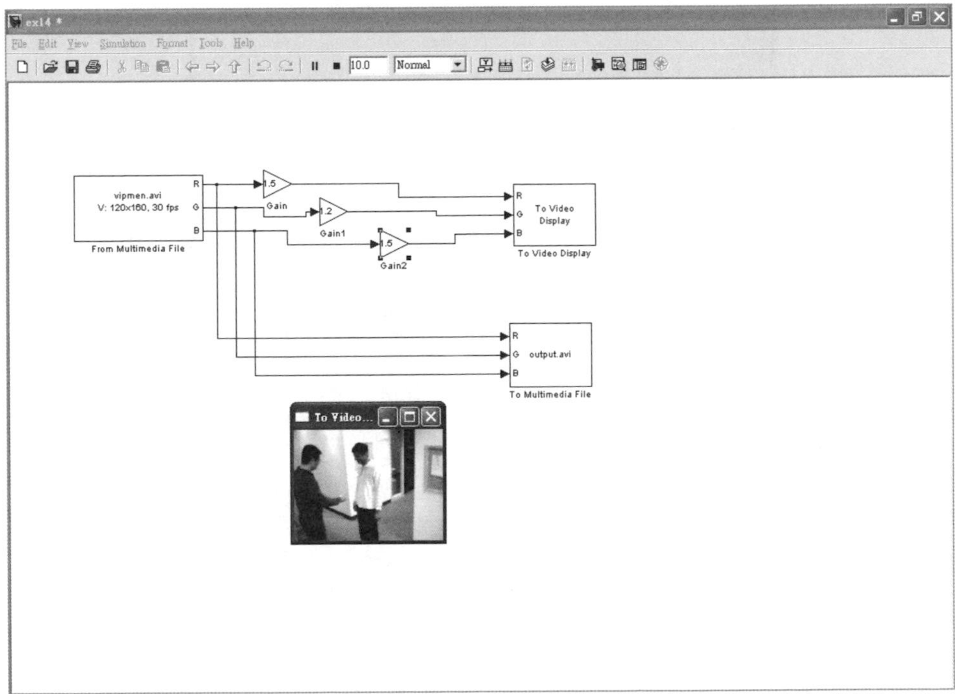
ex14 *
File Edit View Simulation Format Tools Help
10.0
Normal
vipmen.avi
V: 120x160, 30 fps
From Multimedia File
Gain
Gain1
Gain2
To Video Display
To Video Display
output.avi
To Multimedia File
To Video...

2-1 Converting Between Intensity and Binary Images Using Relational Operators (Solution)【2-1 使用關係運算子來轉換灰階與二元值影像(問題解答)】

1. Set “Constant value=150” in block (2), and what will be changed in the output image?【1.如果方塊（2）設定“Constant value(常數值)=150”，則輸出圖像會有什麼改變？】

- Hints(提示)

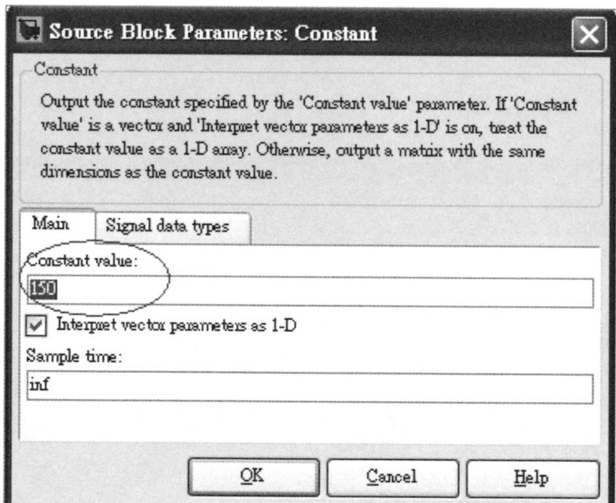

- Results(結果)

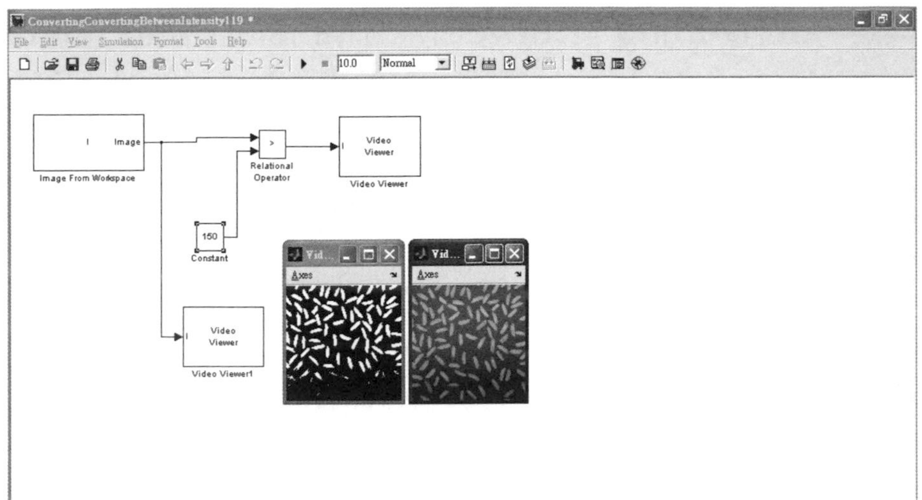
ConvertingConvertingBetweenIntensity119 *
File Edit View Simulation Format Tools Help
10.0
Normal
I
Image
Image From Workspace
>
Relational
Operator
Video
Viewer
Video Viewer
150
Constant
Video
Viewer
Video Viewer1
Axes
Axes

2-2 Converting Between Intensity and Binary Images Using the Auto-threshold Block (Solution)

【2-2 使用自動臨界值法轉換灰階與二元值影像(問題解答)】

1. Select a new picture, and run the program again.
   【1.選擇一個新的圖片，並再次執行該程式。】

- Hints(提示)

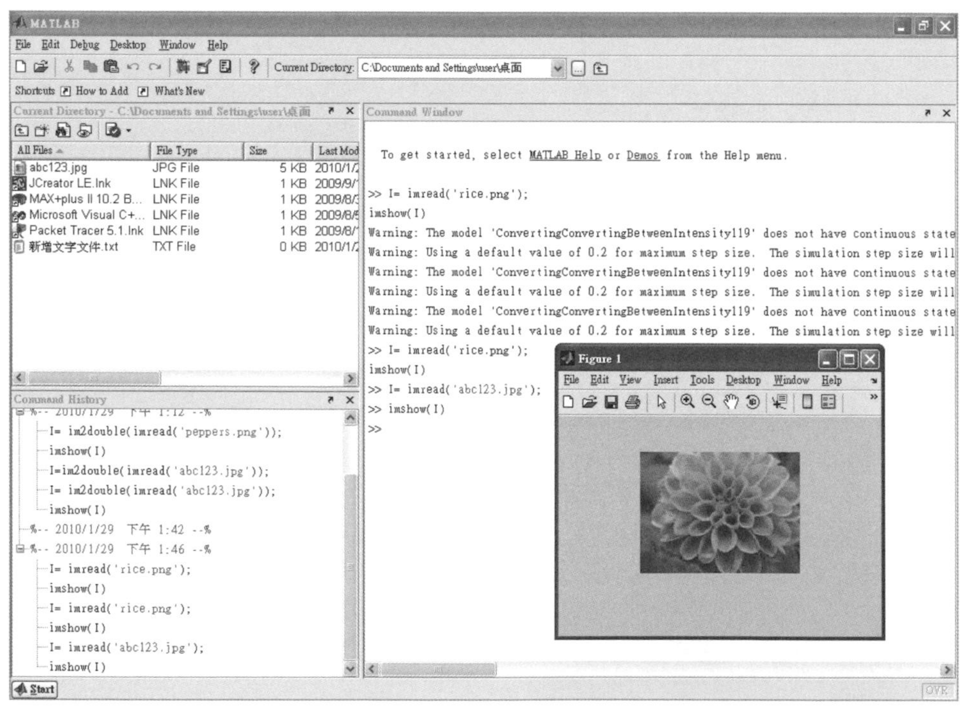

- Results(結果)(1)

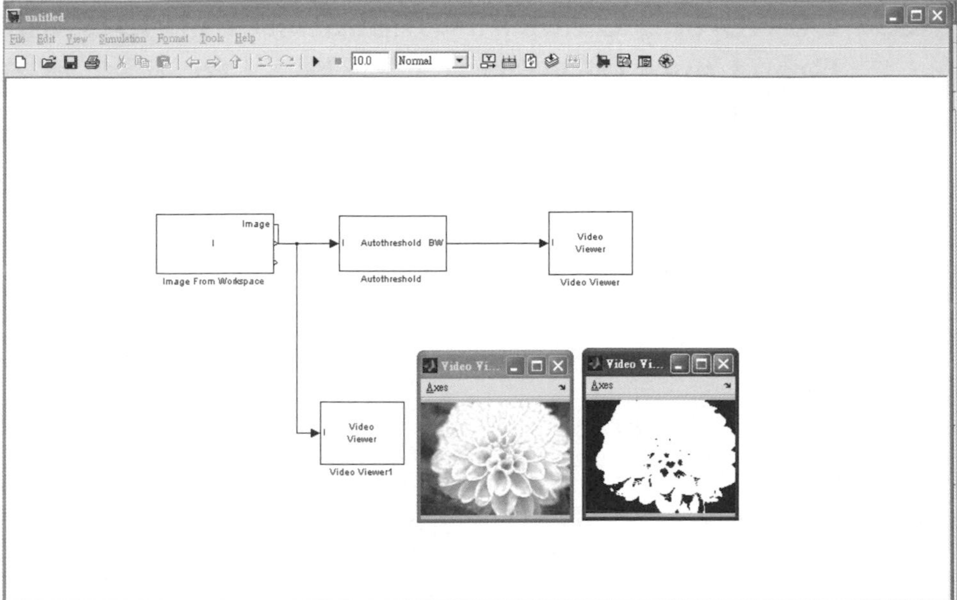

- Results(結果)(2)

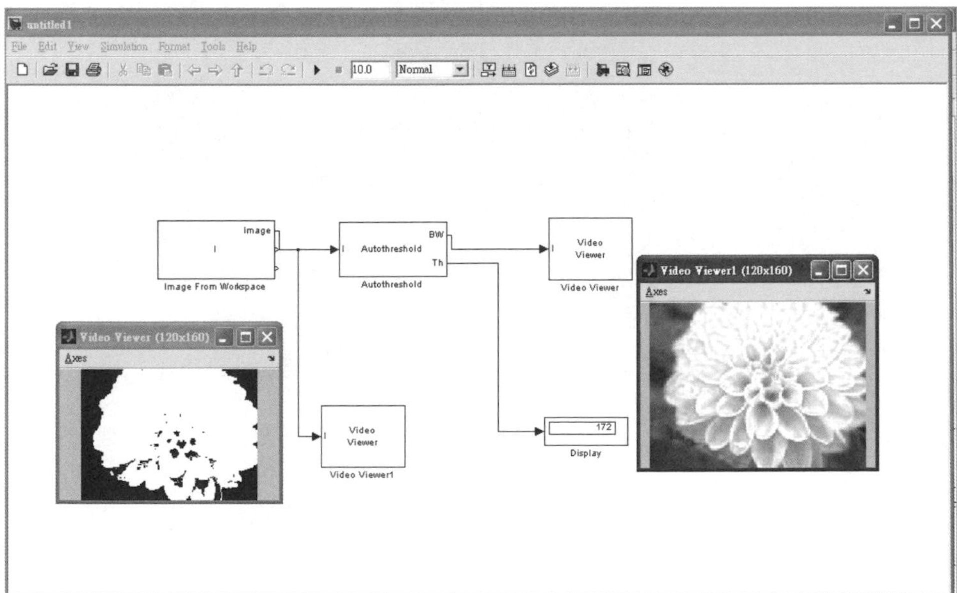

2-3 Converting Color Information from R'G'B' into Gray Level (Solution)【2-3將彩色訊號轉為灰階(問題解答)】

1. Select a new picture, and run the program again.
   【1.選擇一個新的圖片，並再次執行該程式。】

- Hints(提示)

```
I= imread('waterfall.jpg');
imshow(I)
```

- Results(結果)

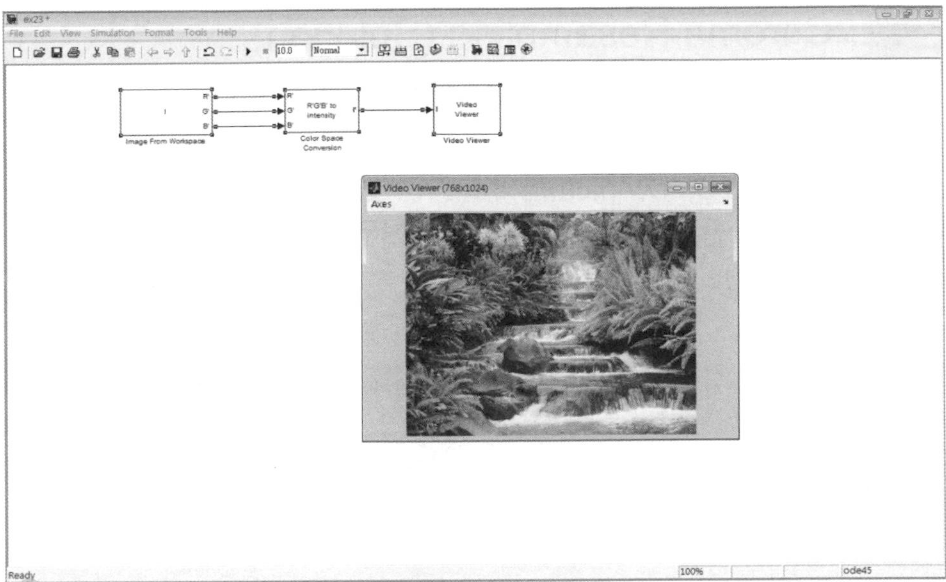
ex23 *
File Edit View Simulation Format Tools Help
10.0
Normal
R'
G'
B'
Image From Workspace
R'G'B' to intensity
I'
Color Space Conversion
Video Viewer
Video Viewer
Video Viewer (768x1024)
Axes
Ready
100%
ode45

2-4 Chroma Resampling (Solution)
【2-4 色度重新取樣(問題解答)】

1. Please select a new picture, and run the program again.
【1.請選擇一個新的圖片，並再次執行該程式。】

- Hints(提示)

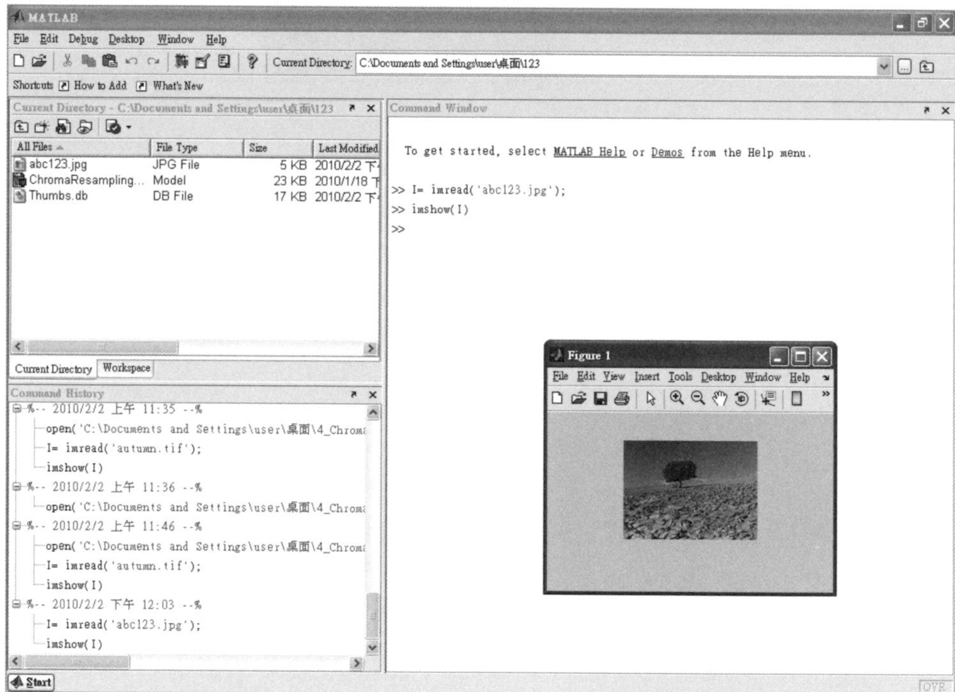

- Results(結果)

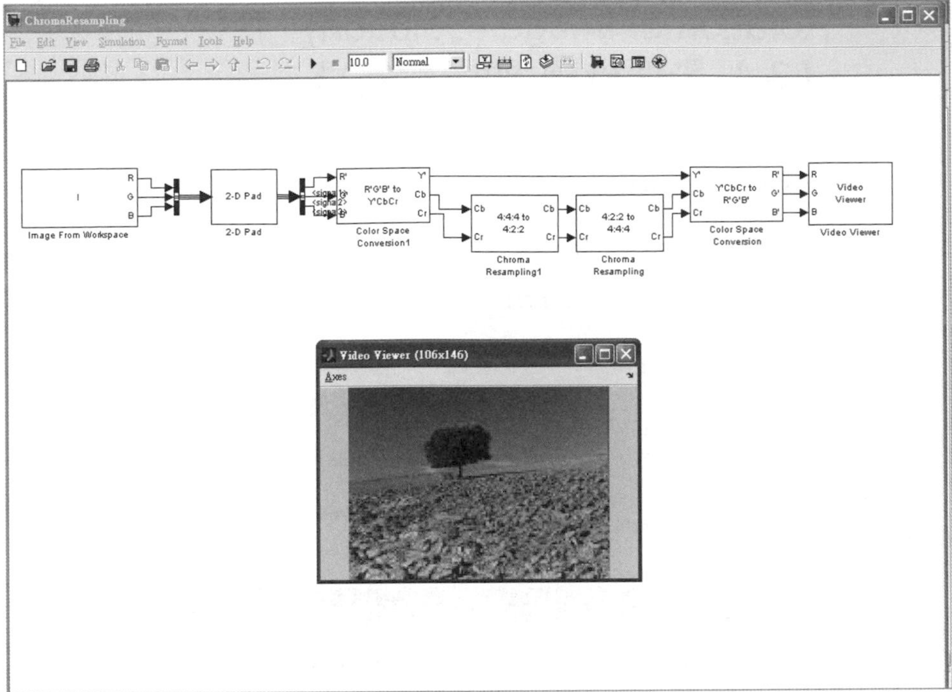
ChromaResampling
File Edit View Simulation Format Tools Help
10.0
Normal
Image From Workspace
2-D Pad
R'G'B' to Y'CbCr
Color Space Conversion1
4:4:4 to 4:2:2
Chroma Resampling1
4:2:2 to 4:4:4
Chroma Resampling
Y'CbCr to R'G'B'
Color Space Conversion
Video Viewer
Video Viewer
Video Viewer (106x146)
Axes

3-1 Rotation of Image (Solution)
【3-1 影像旋轉(問題解答)】

1. Change the parameters in the rotating Video Viewer (Block (3))?

   Interpolation method=Bilinear→Bicubic→Nearest neighbor

   Display rotated image in=center→Top-left Conner

【1.將方塊（3）旋轉器的參數改變如下：

內插法=雙綫性逼近→三次方逼近→參考最鄰近像素

以何種方式旋轉=中心點為旋轉中心→施轉並碰撞角落】

- (a)Hints(提示)

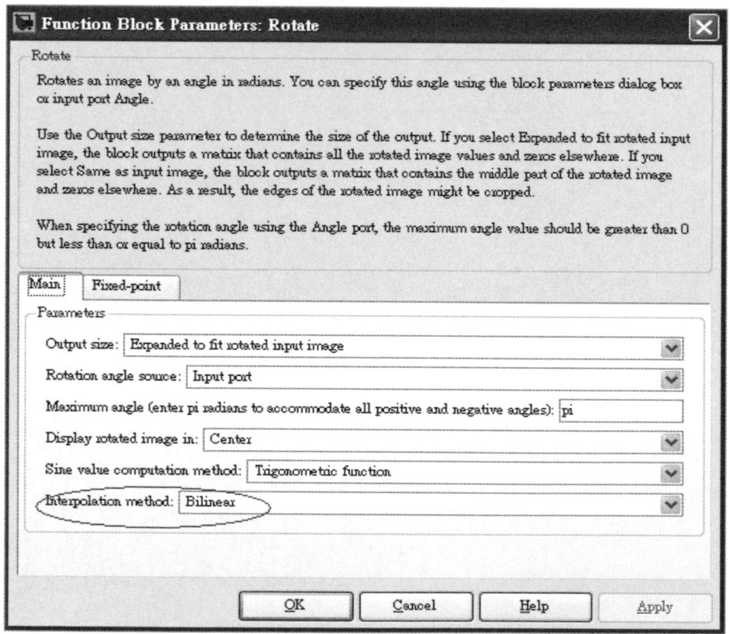

- (a)Result(結果)

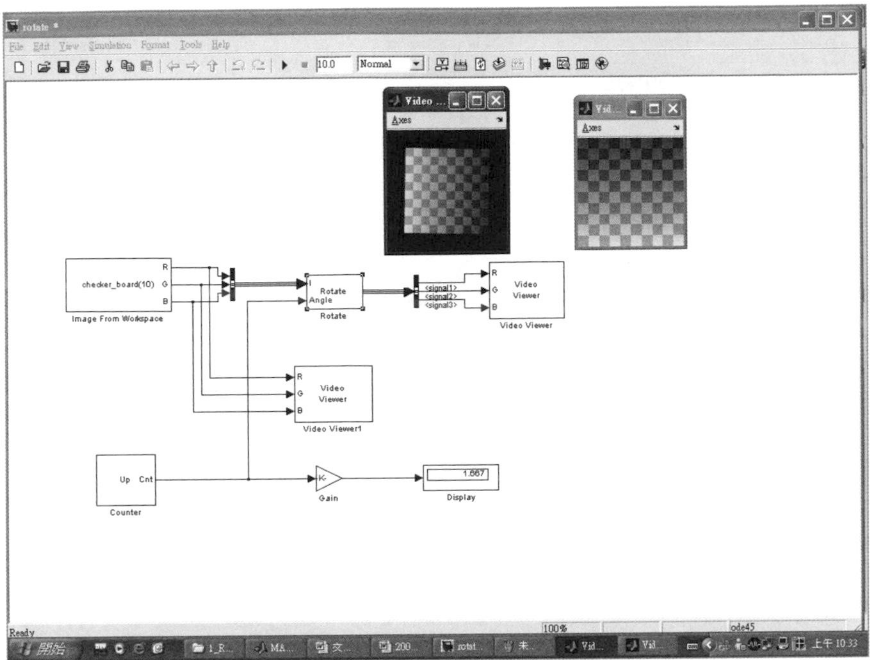

- (a)Hints(提示)

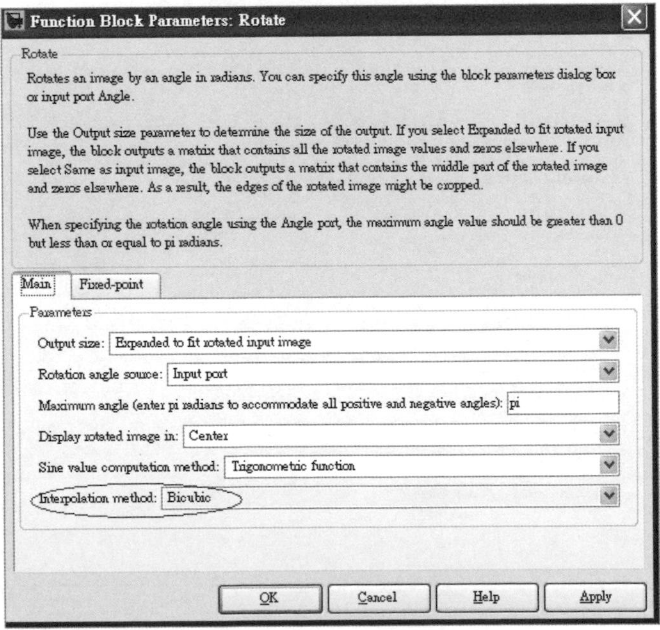

- (a)Result(結果)

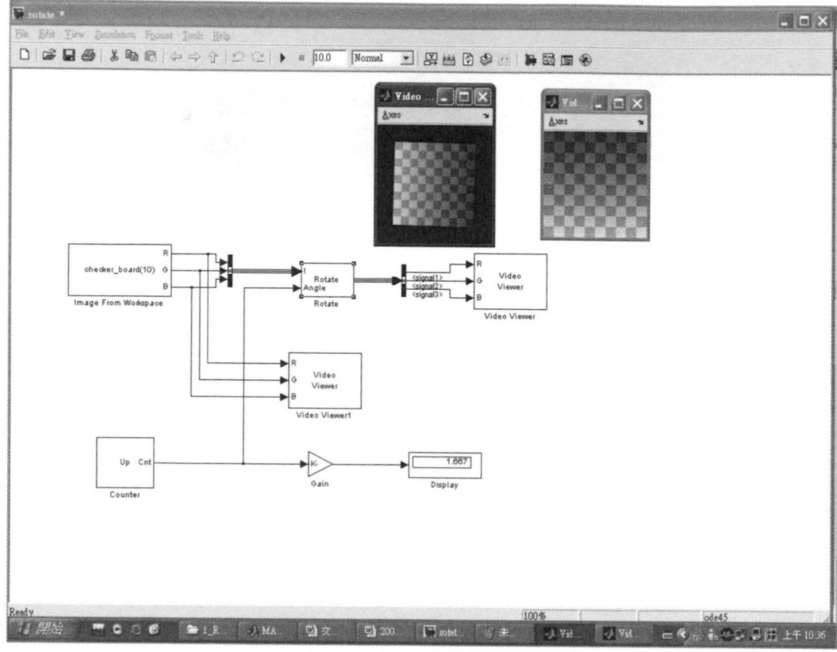

- (a)Hints(提示)

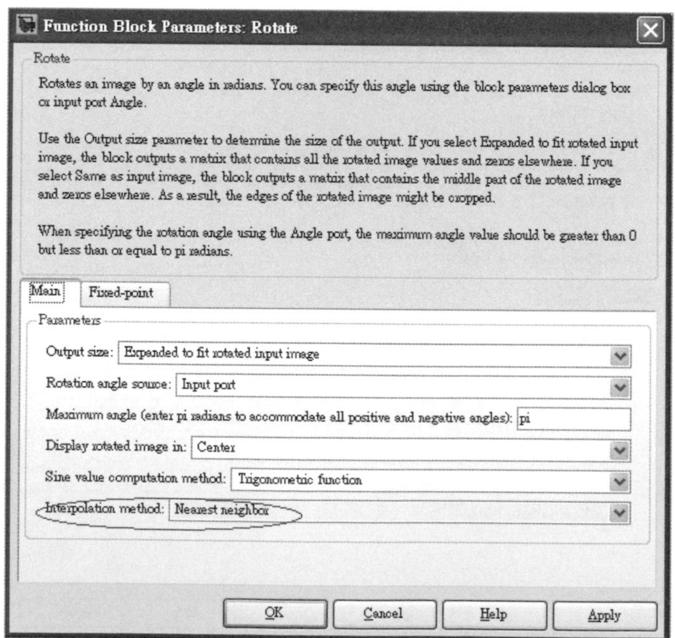

- (a)Result(結果)

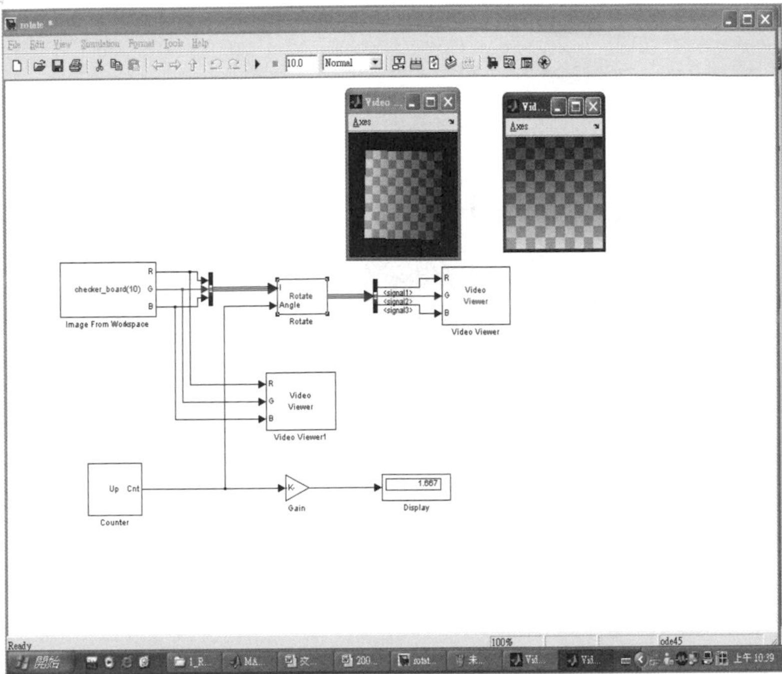

- (b)Hints(提示)

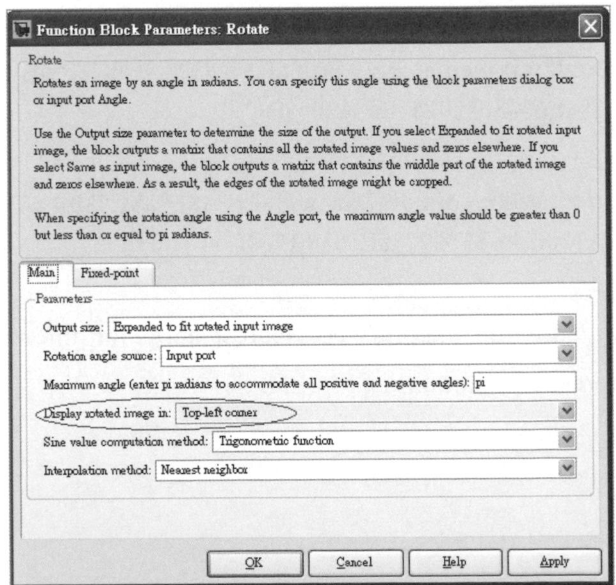

- (b)Result(結果)

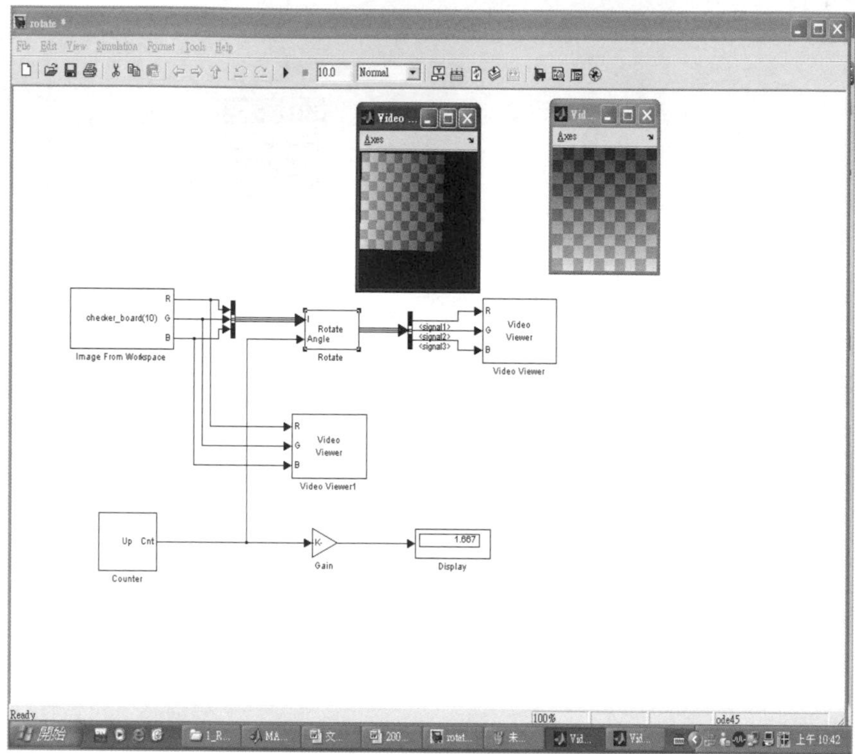

Bilinear:
http://zh.wikipedia.org/wiki/%E5%8F%8C%E7%BA%BF%E6%80%A7%E6%8F%92%E5%80%BC

Bicubic:
http://zh.wikipedia.org/zh-tw/%E5%8F%8C%E4%B8%89%E6%AC%A1%E6%8F%92%E5%80%BC

Nearest neighbor:
http://en.wikipedia.org/wiki/Nearest-neighbor_interpolation

Reference: Examples of MatLab

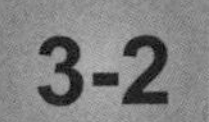

3-2 Resizing the Image (Solution)【3-2 重新調整影像大小(問題解答)】

1. What will be changed in the resizing Video Viewer? Why?

   (a) Resizing factor=10, in Resize block (2)

   (b) Resizing factor=30, in Resize block (2)

   (c) Resizing factor=80, in Resize block (2)

【1.如果參數改變如下，在視頻播放器將有何改變？ 為什麼？

   (a) 在方塊(2)中，Resizing factor(重新調整大小因子)=10

   (b) 在方塊(2)中，Resizing factor(重新調整大小因子)=30

   (c) 在方塊(2)中，Resizing factor(重新調整大小因子)=80】

- (a)Hints(提示)

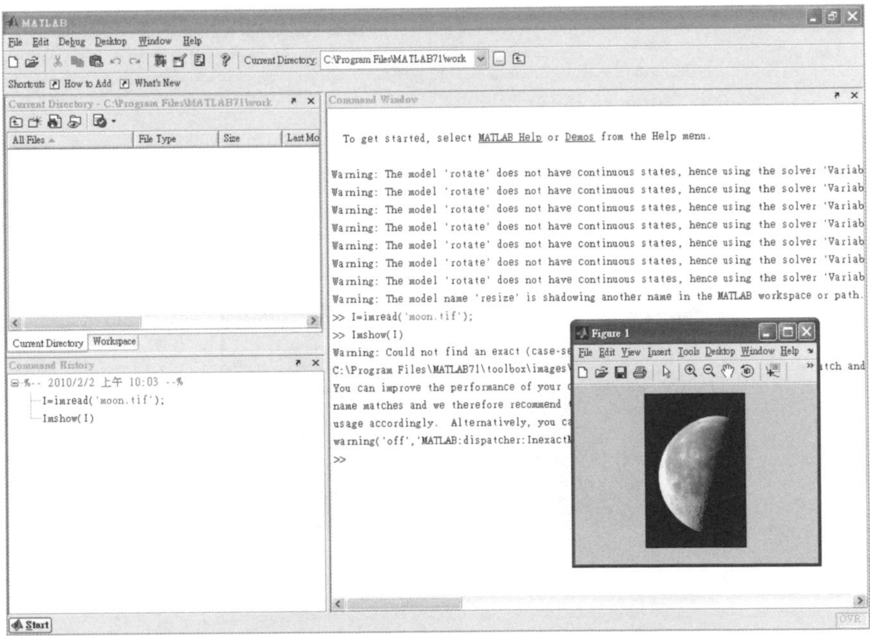

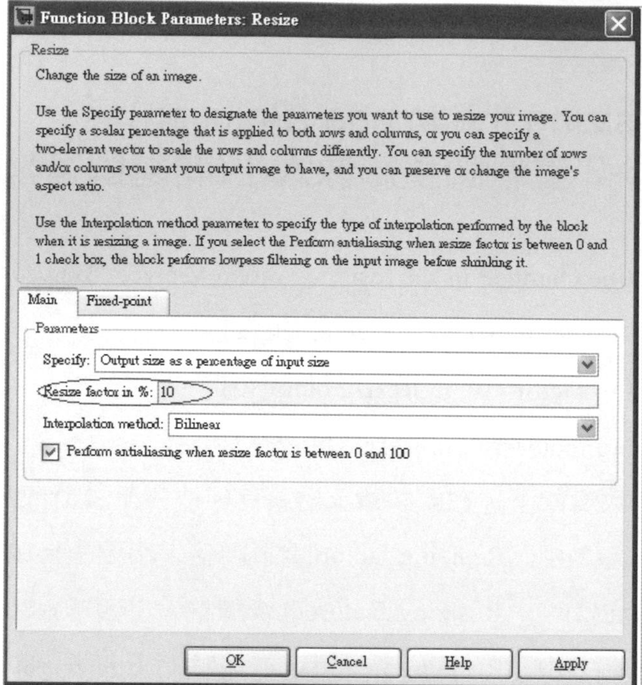
Function Block Parameters: Resize
Resize
Change the size of an image.
Use the Specify parameter to designate the parameters you want to use to resize your image. You can specify a scalar percentage that is applied to both rows and columns, or you can specify a two-element vector to scale the rows and columns differently. You can specify the number of rows and/or columns you want your output image to have, and you can preserve or change the image's aspect ratio.
Use the Interpolation method parameter to specify the type of interpolation performed by the block when it is resizing a image. If you select the Perform antialiasing when resize factor is between 0 and 1 check box, the block performs lowpass filtering on the input image before shrinking it.
Main
Fixed-point
Parameters
Specify: Output size as a percentage of input size
Resize factor in %: 10
Interpolation method: Bilinear
Perform antialiasing when resize factor is between 0 and 100
OK
Cancel
Help
Apply

- (a)Result(結果)

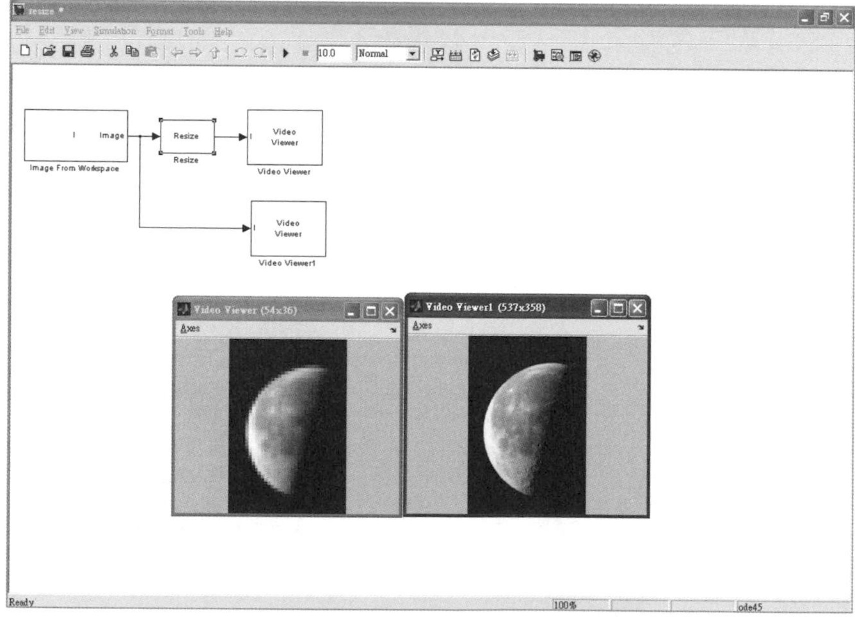
Image
Image From Workspace
Resize
Video Viewer
Video Viewer1
Video Viewer (54x36)
Video Viewer1 (537x358)
Ready
100%
ode45

- (b)Hints(提示)

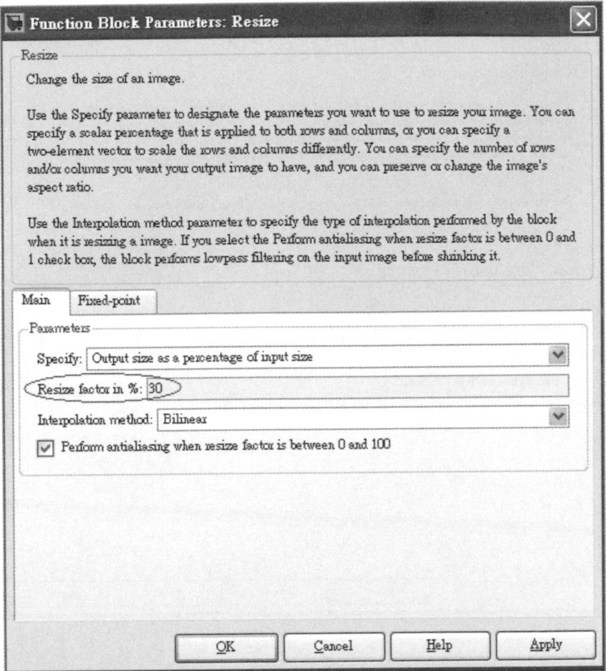

- (b)Results(結果)

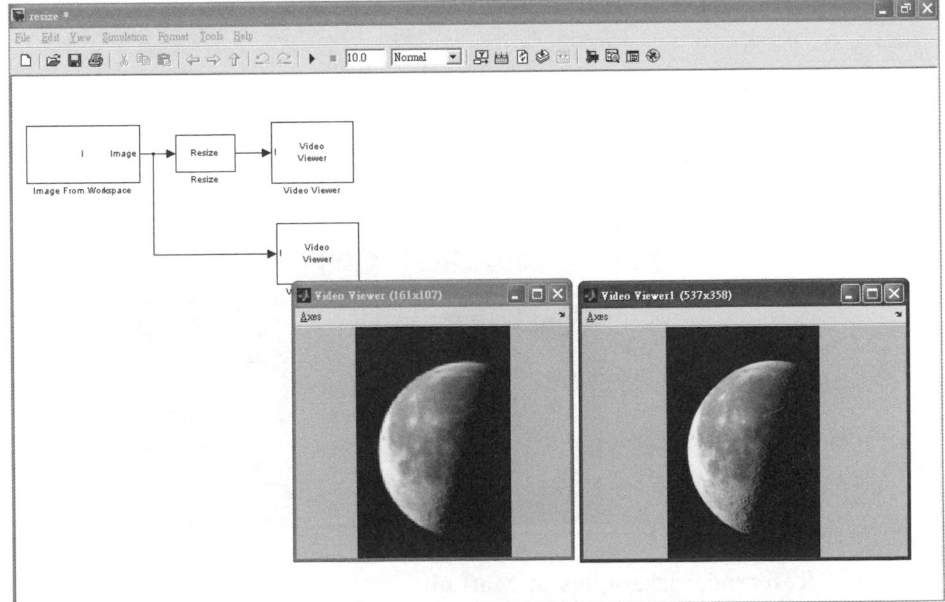

- (c)Hints(提示)

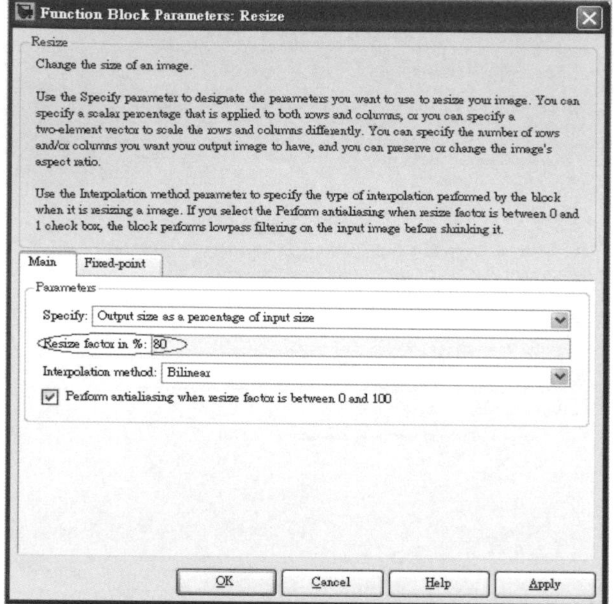

- (c)Hints(結果)

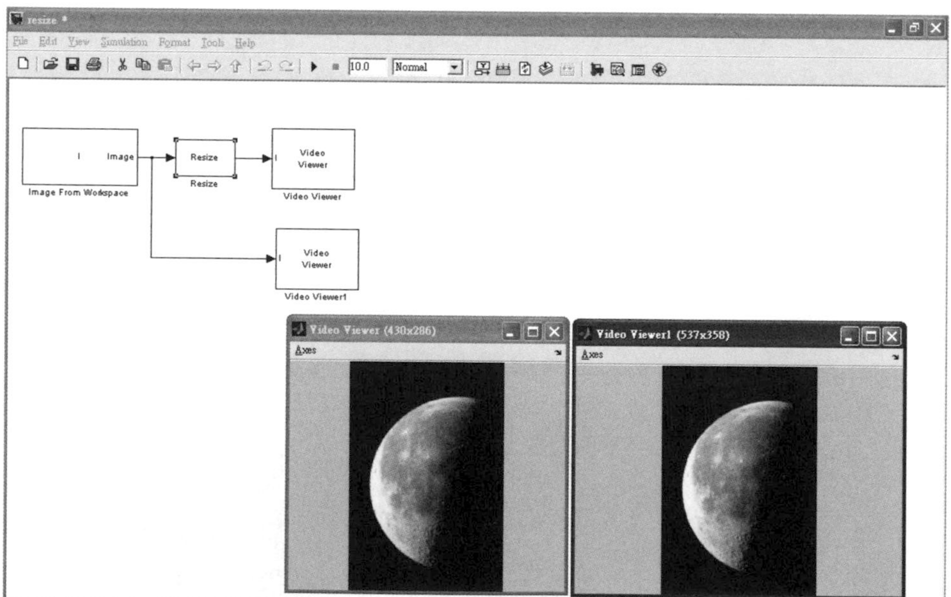

Reference: Examples of MatLab

3-3 Cropping the Image (Solution)【3-3 剪輯影像(問題解答)】

1. What is the output, if the settings are changed as follows? Why?

   (a) Rows = 140→1; Columns = 200→115;

   Output port dimensions = [65 65]

   (b) Rows = 90; Columns = 140;

   Output port dimensions = [65 65]

【1.如果設定值改變為以下數值則輸出有何改變?為什麼？

   (a) Rows 行=由 140 改為 1, Columns 列=由 200 改為 115;

   Output port dimensions 輸出阜尺寸= [65 65]

   (b) Rows 行= 90; Columns 列= 140;

   Output port dimensions 輸出阜尺寸= [65 65]】

2. Please crop a coin.

   【2.請取出一個完整的硬幣。】

- 1.(a)Hints(提示)

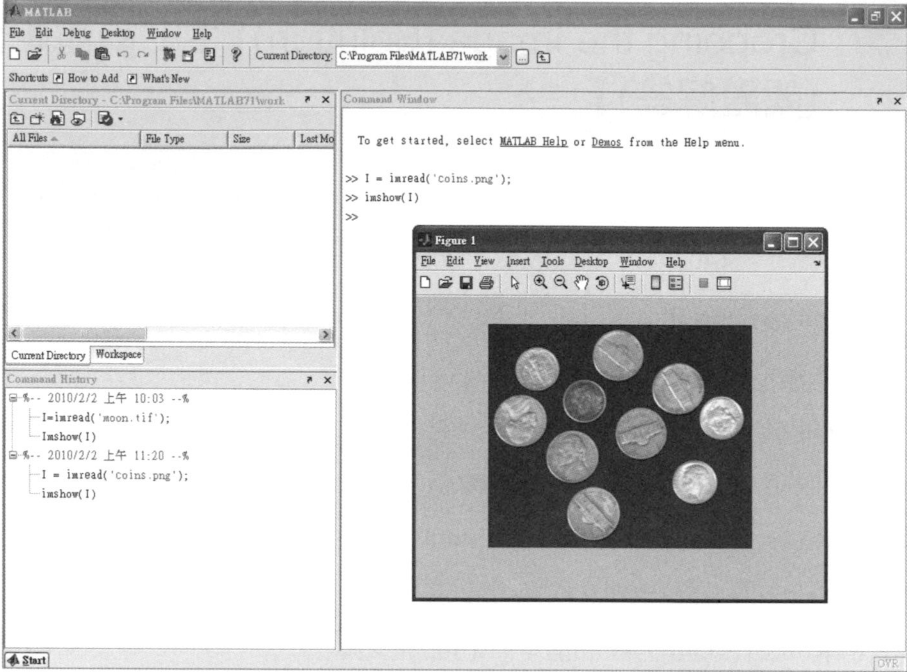
MATLAB
File Edit Debug Desktop Window Help
Current Directory: C:\Program Files\MATLAB71\work
Shortcuts How to Add What's New
Current Directory - C:\Program Files\MATLAB71\work
All Files File Type Size Last Mo
Current Directory Workspace
Command History
%-- 2010/2/2 上午 10:03 --%
I=imread('moon.tif');
Imshow(I)
%-- 2010/2/2 上午 11:20 --%
I = imread('coins.png');
imshow(I)
Command Window
To get started, select MATLAB Help or Demos from the Help menu.
>> I = imread('coins.png');
>> imshow(I)
>>
Figure 1
File Edit View Insert Tools Desktop Window Help
Start
OVR

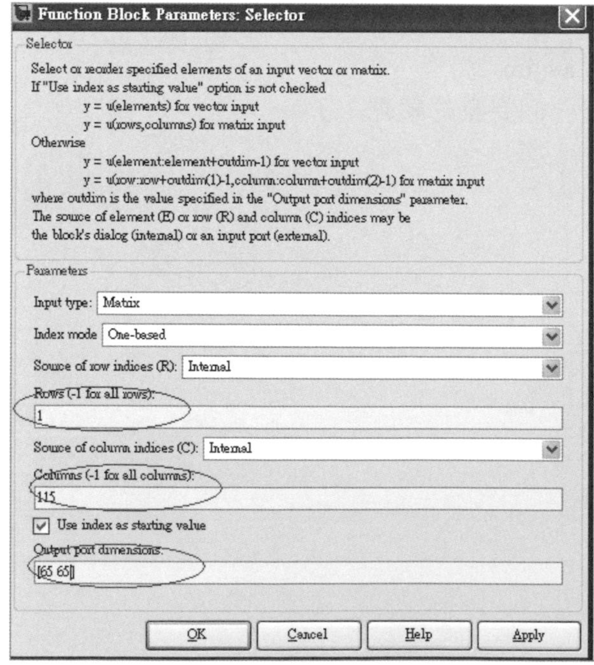
Function Block Parameters: Selector
Selector
Select or reorder specified elements of an input vector or matrix.
If "Use index as starting value" option is not checked
y = u(elements) for vector input
y = u(rows,columns) for matrix input
Otherwise
y = u(element:element+outdim-1) for vector input
y = u(row:row+outdim(1)-1,column:column+outdim(2)-1) for matrix input
where outdim is the value specified in the "Output port dimensions" parameter.
The source of element (E) or row (R) and column (C) indices may be
the block's dialog (internal) or an input port (external).
Parameters
Input type: Matrix
Index mode One-based
Source of row indices (R): Internal
Rows (-1 for all rows):
1
Source of column indices (C): Internal
Columns (-1 for all columns):
115
Use index as starting value
Output port dimensions:
[65 65]
OK
Cancel
Help
Apply

- (a)Results(結果)

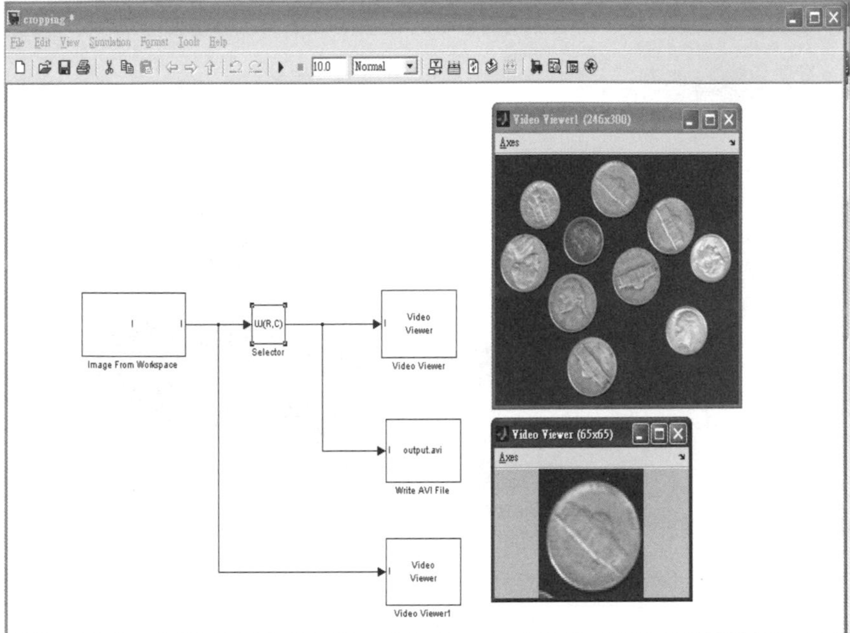

- (b)Hints(提示)

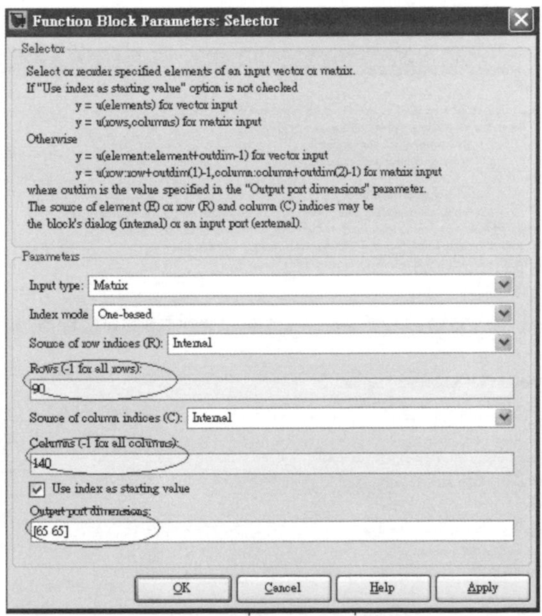

- (b)Results(結果)

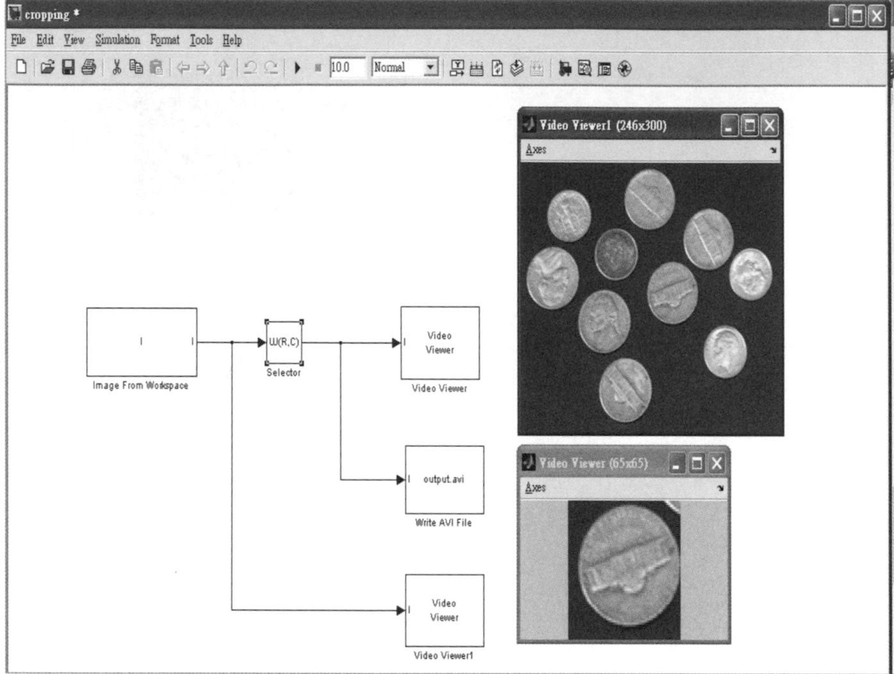

- 2.Hint(提示)

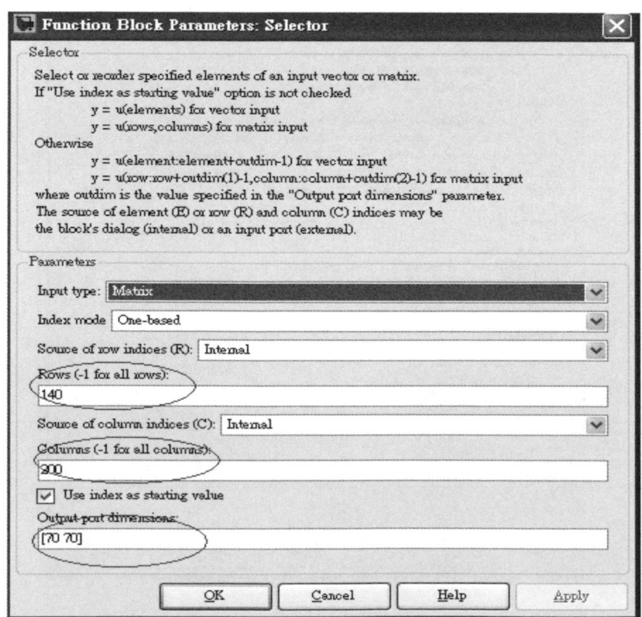

- Results(結果)

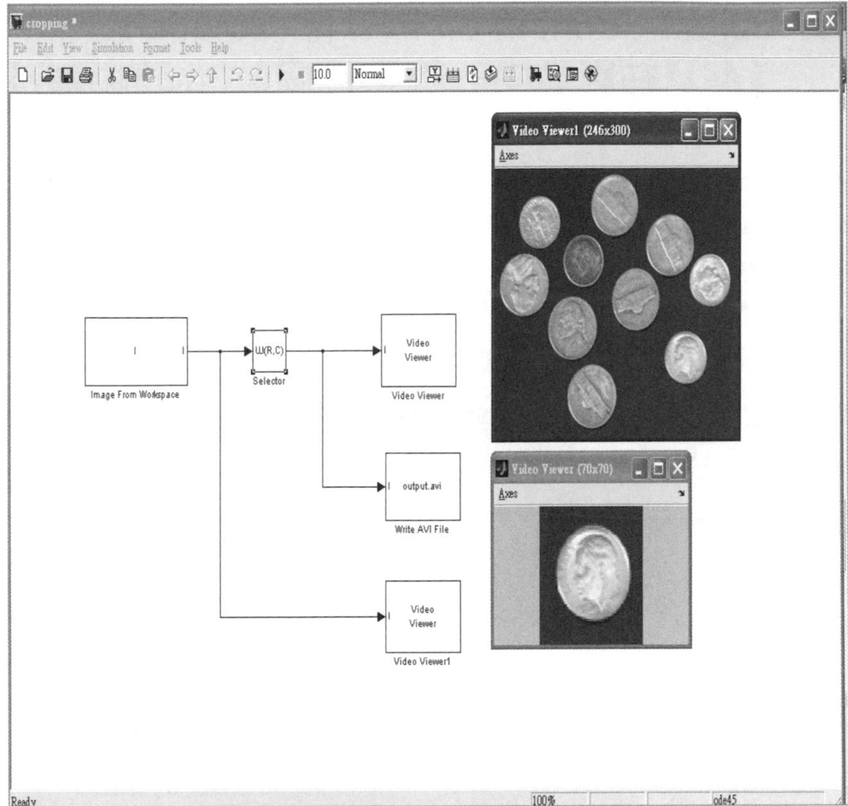

Reference: Examples of MatLab

4-1 Object Count in an Image (Solution)【4-1 在影像中計數物件個數(問題解答)】

1. Set “Constant value =250” in block (3), and find the changes in the output.【1.將方塊(3)的 Constant value 設為 250，並觀察其輸出的改變。】

- Hints(提示)

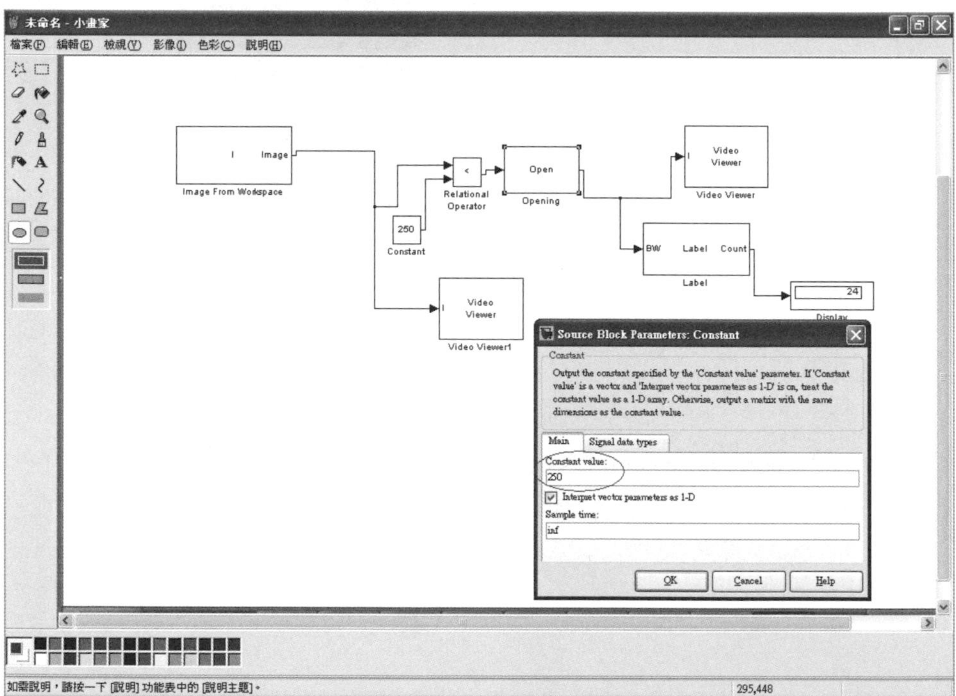

- Results(結果)

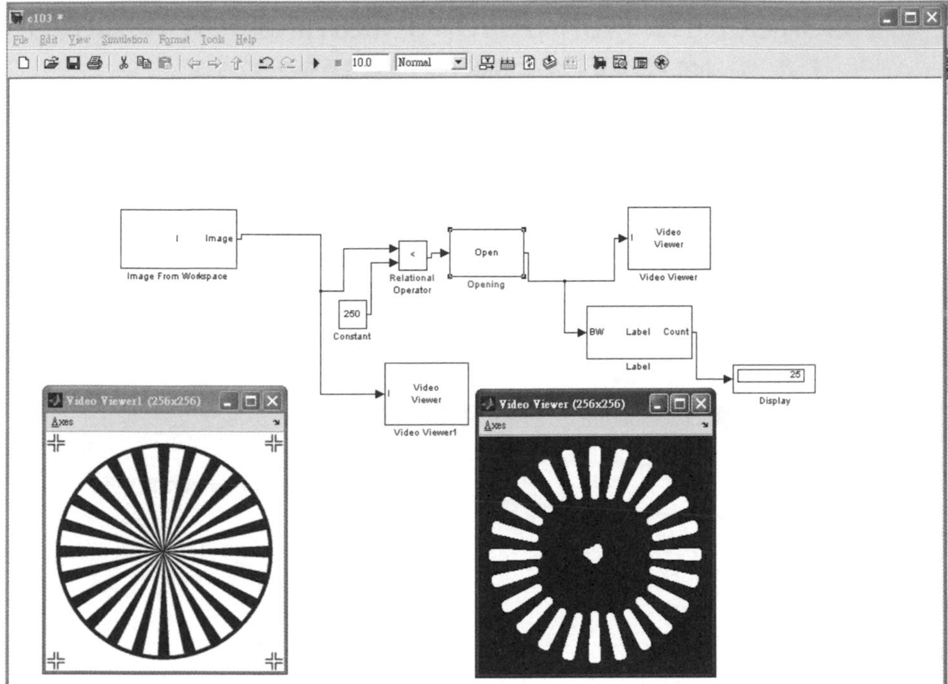
c103 *
File Edit View Simulation Format Tools Help
10.0
Normal
I Image
Image From Workspace
<
Relational Operator
250
Constant
Open
Opening
Video Viewer
Video Viewer
BW Label Count
Label
25
Display
Video Viewer
Video Viewer1
Video Viewer1 (256x256)
Axes
Video Viewer (256x256)
Axes

4-2 Correct the Non-uniform Illumination (Solution)【4-2 修正非均勻照度(問題解答)】

1. Set “Constant value = 10, 50, and 70” in block (7), what will be changed in the output image?【1.設定方塊（7）中“Constant value(常數值) = 10, 50, and 70”， 輸出圖像將有什麼改變？】

- Hints(提示)

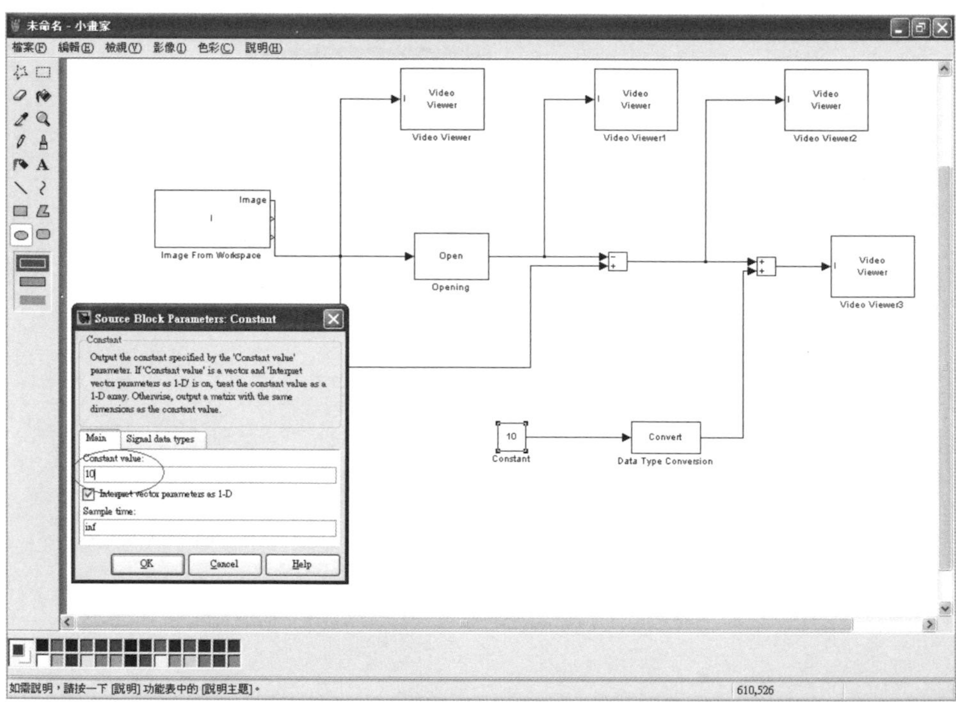

- Results(結果)

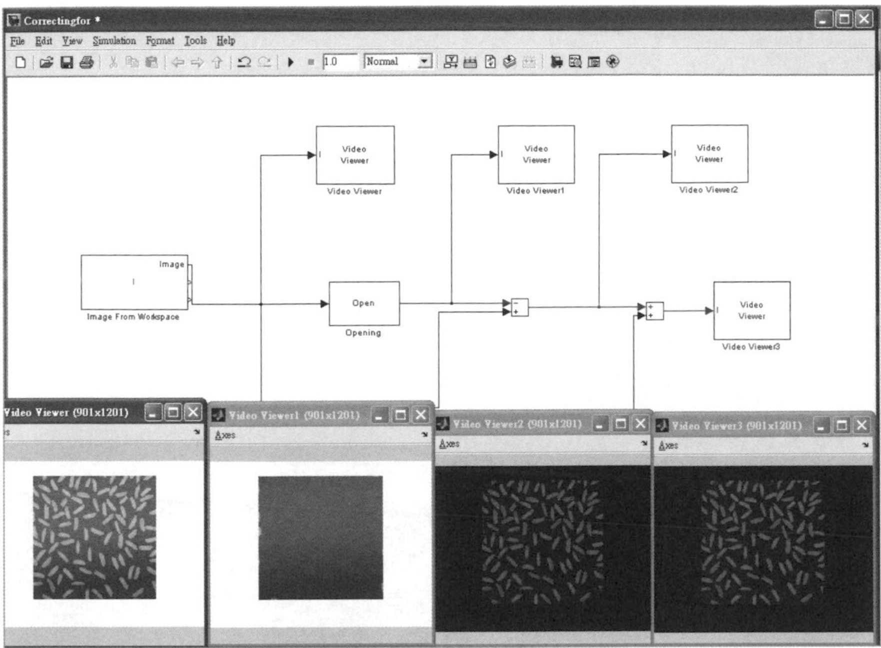

- Hints(提示)

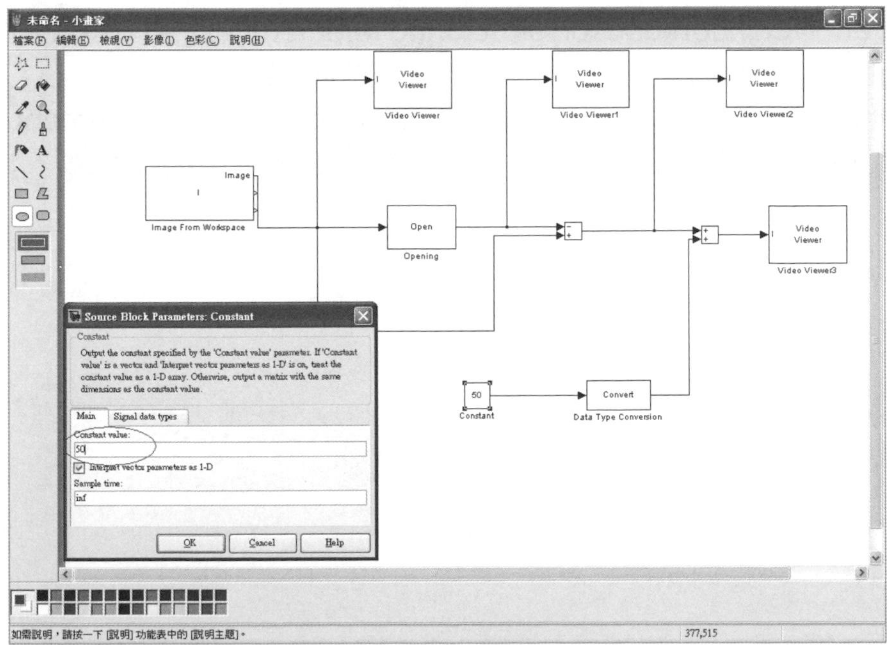

• Results(結果)

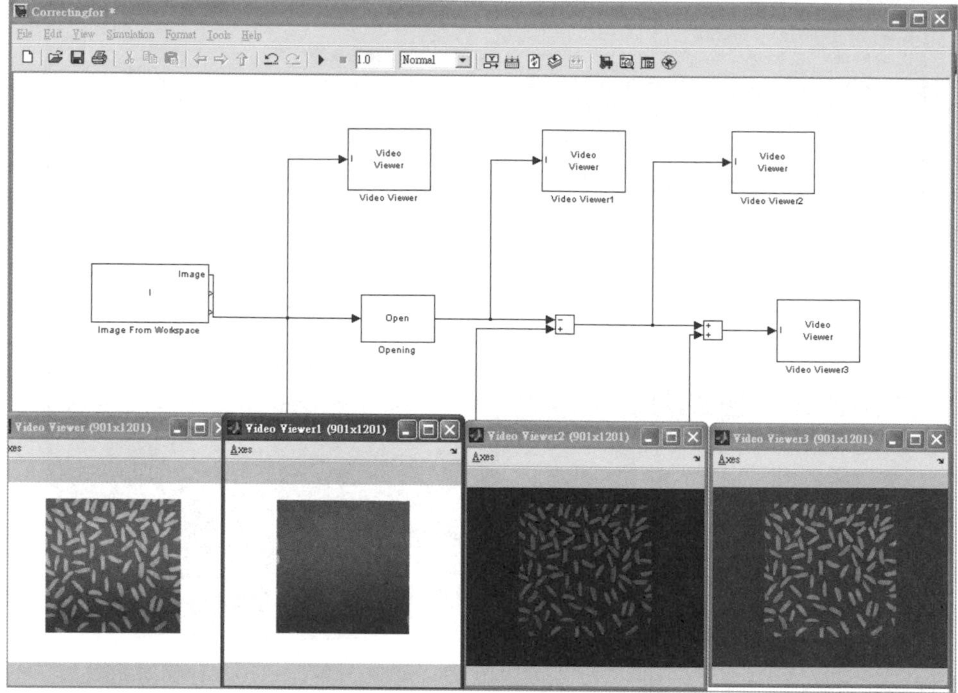

2. Set "Neighborhood or structuring element =strel('disk',30)" in block (2), how will the output images be changed?
【2.設定方塊（2）中"Neighborhood or structuring element (相鄰或結構元素)=strel('disk',30)"， 輸出圖像將會如何改變？】

- Hints(提示)

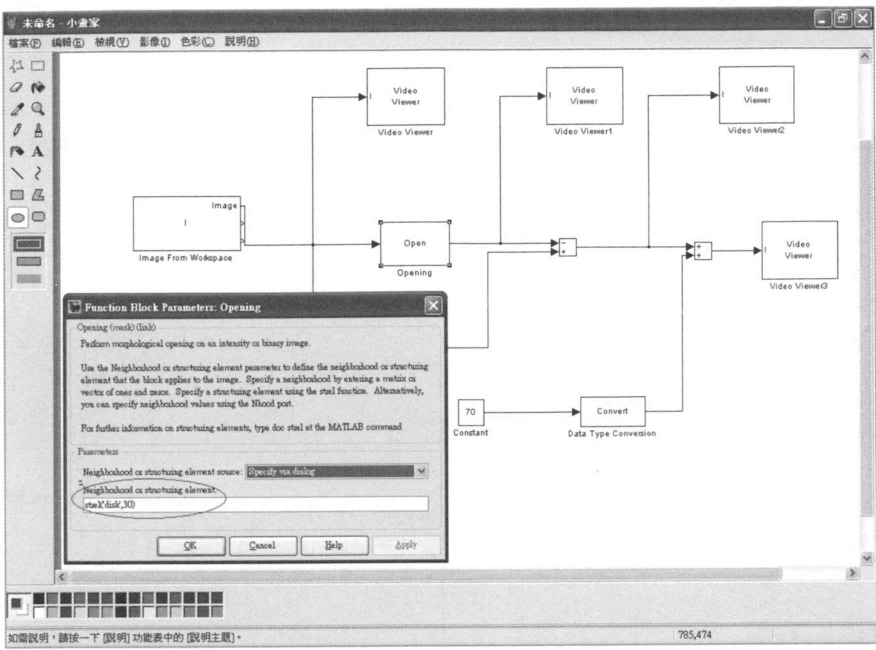

- Results(結果)

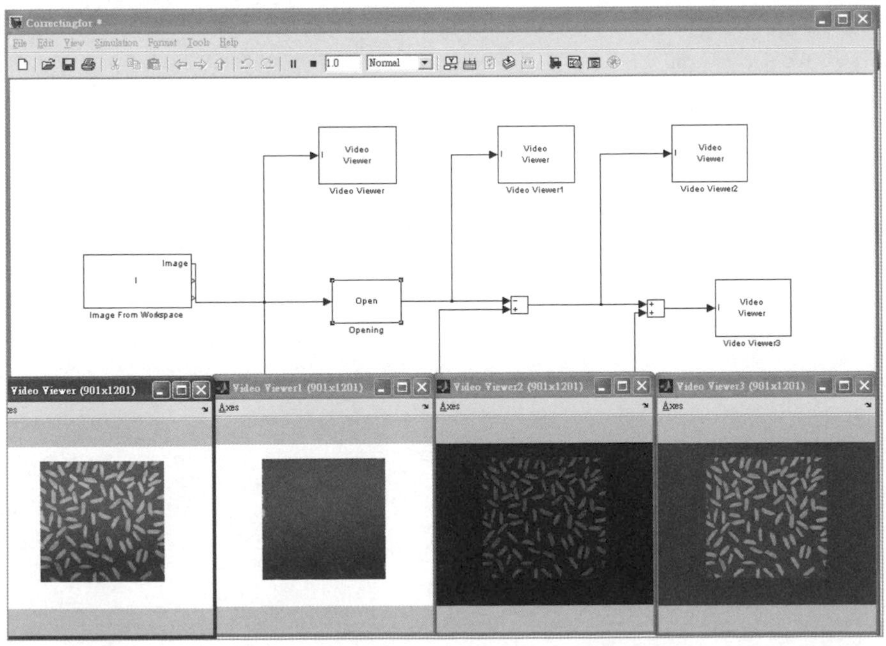

3. Select color map in the viewers of blocks (3) to (6).【3.選擇色彩量表在放映方塊（3）至（6）中。】

- Hints(提示)

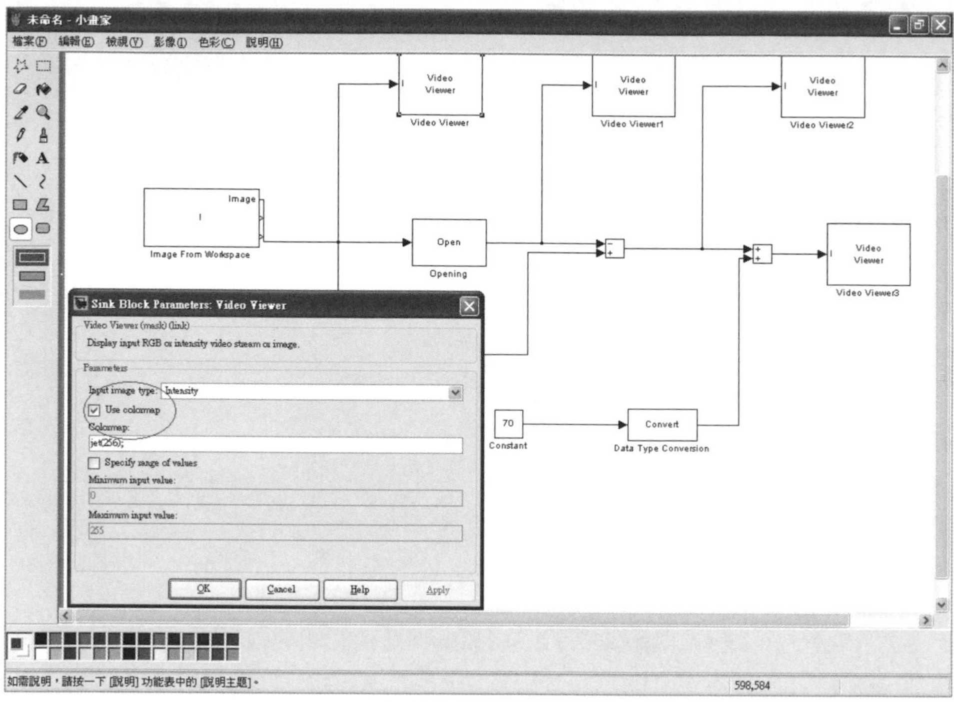

- Results(結果)

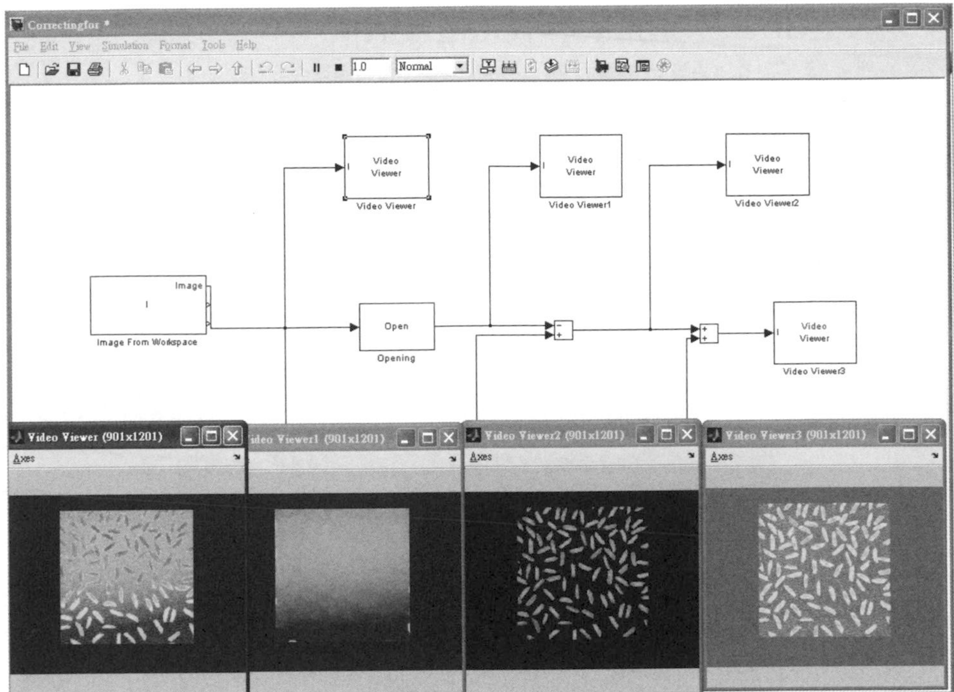
Correctingfor *
File Edit View Simulation Format Tools Help
Normal
Video Viewer
Video Viewer
Video Viewer1
Video Viewer2
Image
Image From Workspace
Open
Opening
Video Viewer3
Video Viewer (901x1201)
Video Viewer1 (901x1201)
Video Viewer2 (901x1201)
Video Viewer3 (901x1201)
Axes

5-1 Edge Detection (Solution)
【5-1 偵測圖像的邊緣(問題解答)】

1. What will be changed in the output, if we clear the check box of "Main Edge Thinning" in Sobel block (2)?
   【1.如果我們不點選 Sobel(索貝爾)方塊（2）中的"Main Edge Thinning(主邊緣窄化)"，則輸出會有何改變？】

- Hints(提示)

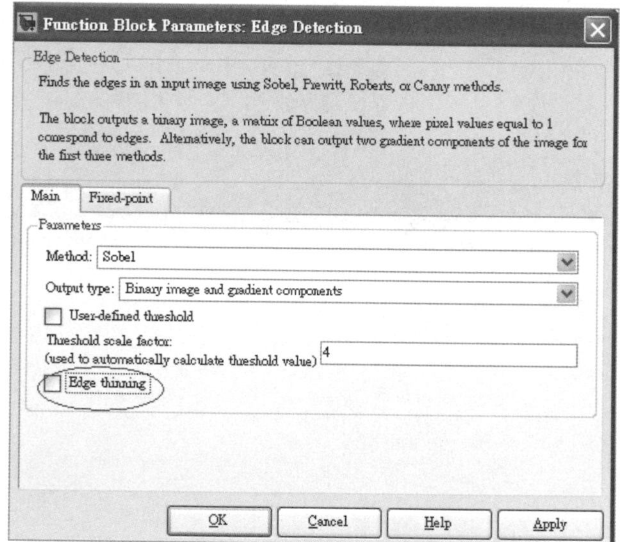

- Results(結果)

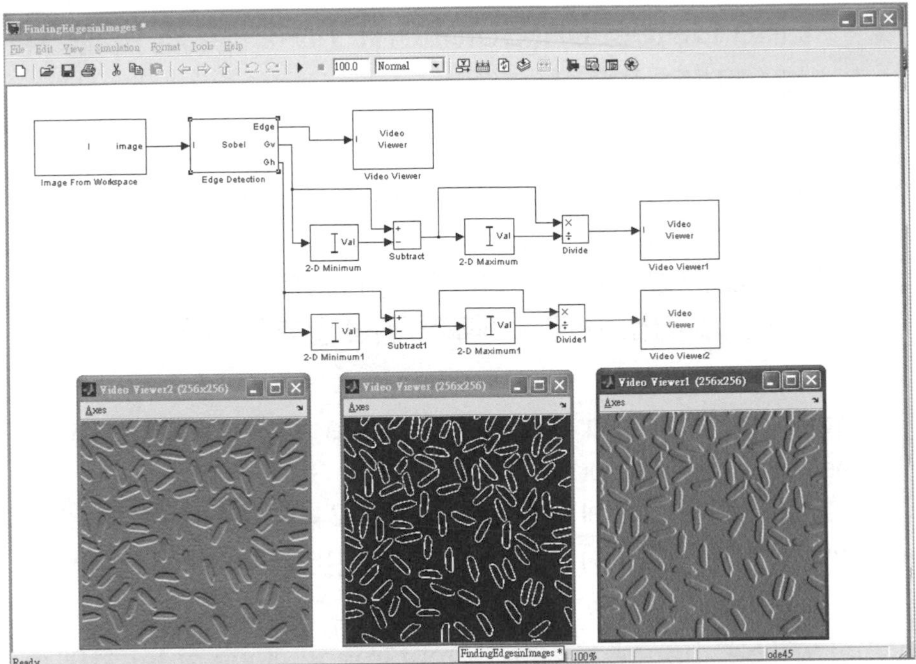
FindingEdgesinImages *
File Edit View Simulation Format Tools Help
100.0
Normal
image
Image From Workspace
Sobel
Edge
Gv
Gh
Edge Detection
Video Viewer
Video Viewer
Val
2-D Minimum
Subtract
2-D Maximum
Divide
Video Viewer1
2-D Minimum1
Subtract1
2-D Maximum1
Divide1
Video Viewer2
Video Viewer2 (256x256)
Video Viewer (256x256)
Video Viewer1 (256x256)
Axes
Ready
100%
ode45

5-2 Line Detection(Solution)【5-2 偵測影像中的線(問題討論)】

1. What will be changed in the output image, if we select the check box of "Edge Thinning" in the block of "Sobel" (2)?
【1. 如果我們在"Sobel"（索貝爾）方塊（2）中點選"Edge Thinning(邊緣窄化)"功能，則輸出圖像會有何變化？】

2. What will be changed in the output image, if we enter "3" in the "threshold scale factor" in the block of "Sobel" (2)?
【2.如果我們在"Sobel" (索貝爾)方塊(2)中，將"Threshold Scale Factor (門檻比例因子) 改為"3"， 則輸出圖像會有何變化？】

3. What will be changed in the output image, if we enter "10" in the threshold in the block of the "Find Local Maxima" (5)?
【3.如果我們將方塊(5)"Find Local Maxima (尋找局部最大值)" 中的 "Threshold(臨界值)"改為 10，則輸出圖像會有何變化？】

- Hints(提示)

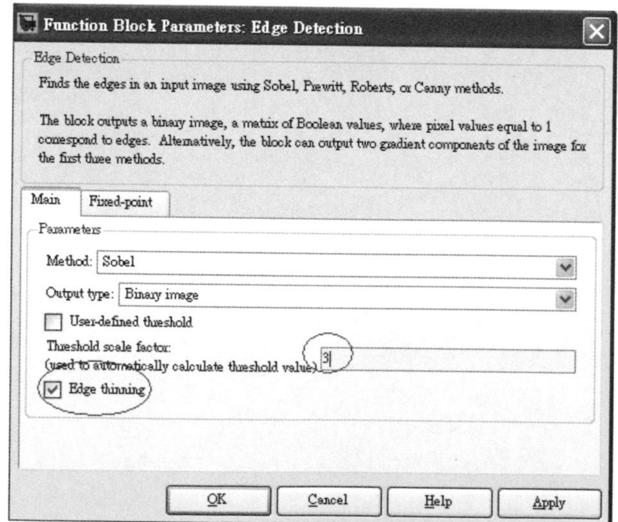

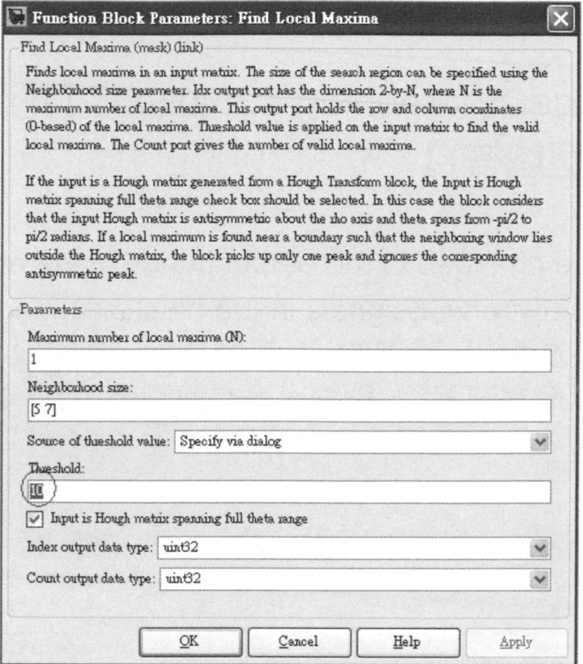
Function Block Parameters: Find Local Maxima
Find Local Maxima (mask) (link)
Finds local maxima in an input matrix. The size of the search region can be specified using the Neighborhood size parameter. Idx output port has the dimension 2-by-N, where N is the maximum number of local maxima. This output port holds the row and column coordinates (0-based) of the local maxima. Threshold value is applied on the input matrix to find the valid local maxima. The Count port gives the number of valid local maxima.
If the input is a Hough matrix generated from a Hough Transform block, the Input is Hough matrix spanning full theta range check box should be selected. In this case the block considers that the input Hough matrix is antisymmetric about the rho axis and theta spans from -pi/2 to pi/2 radians. If a local maximum is found near a boundary such that the neighboring window lies outside the Hough matrix, the block picks up only one peak and ignores the corresponding antisymmetric peak.
Parameters
Maximum number of local maxima (N):
1
Neighborhood size:
[5 7]
Source of threshold value: Specify via dialog
Threshold:
10
Input is Hough matrix spanning full theta range
Index output data type: uint32
Count output data type: uint32
OK
Cancel
Help
Apply

- Results(結果)

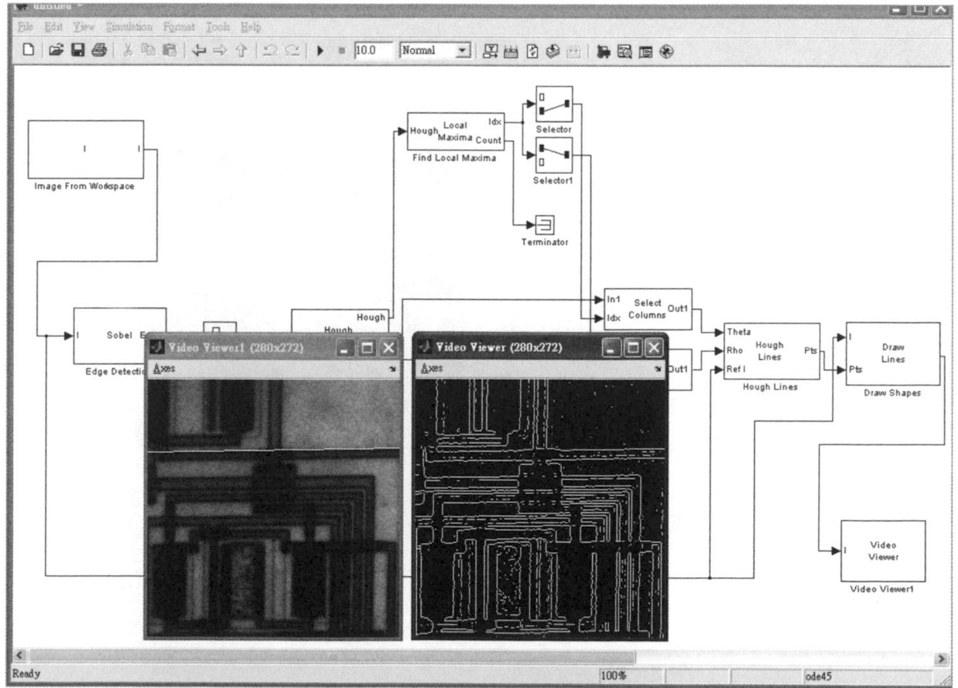
File Edit View Simulation Format Tools Help
10.0
Normal
Image From Workspace
Hough
Local Maxima
Idx
Count
Find Local Maxima
Selector
Selector1
Terminator
Sobel
Edge Detection
Hough
In1
Idx
Select Columns
Out1
Theta
Rho
Ref I
Hough Lines
Pts
Hough Lines
Draw Lines
Draw Shapes
Video Viewer1 (280x272)
Video Viewer (280x272)
Axes
Video Viewer
Video Viewer1
Ready
100%
ode45

6-1 Image Sharpening (Solution)【6-1 影像銳化(問題解答)】

1. What will be changed in the output image, if we add a new “2-D FIR filter” between Cr signals in the blocks (2) and (4)?
【1.如果在方塊(2) 與(4)的 Cr 訊號加上一個“2-D FIR filter(二維無限脈衝響應濾波器)”，則輸出會有何變化？】

- Hints(提示)

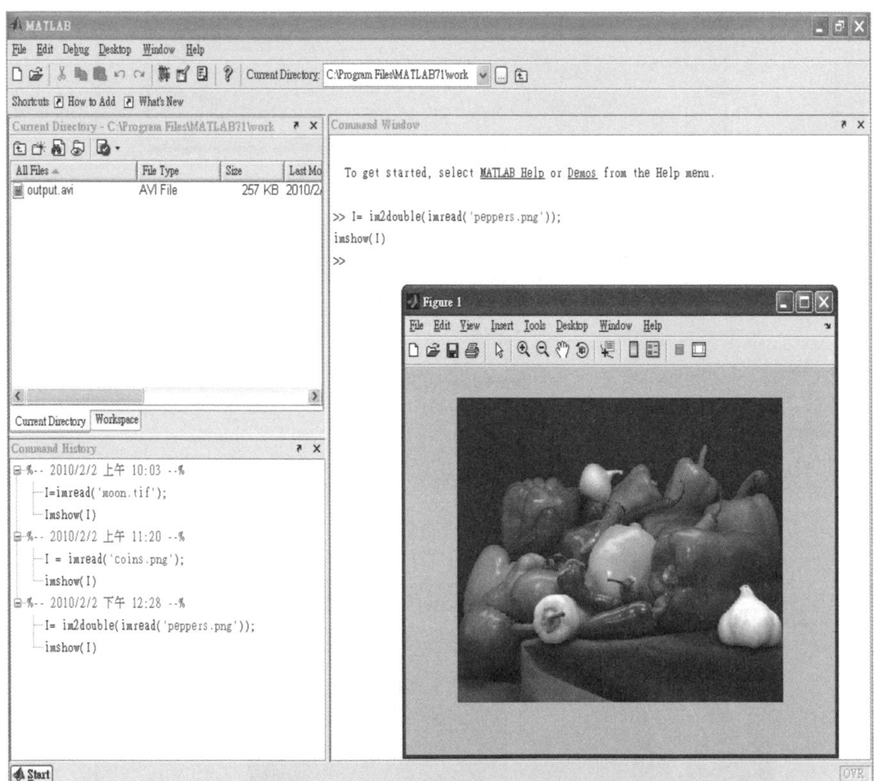

- Results(結果)

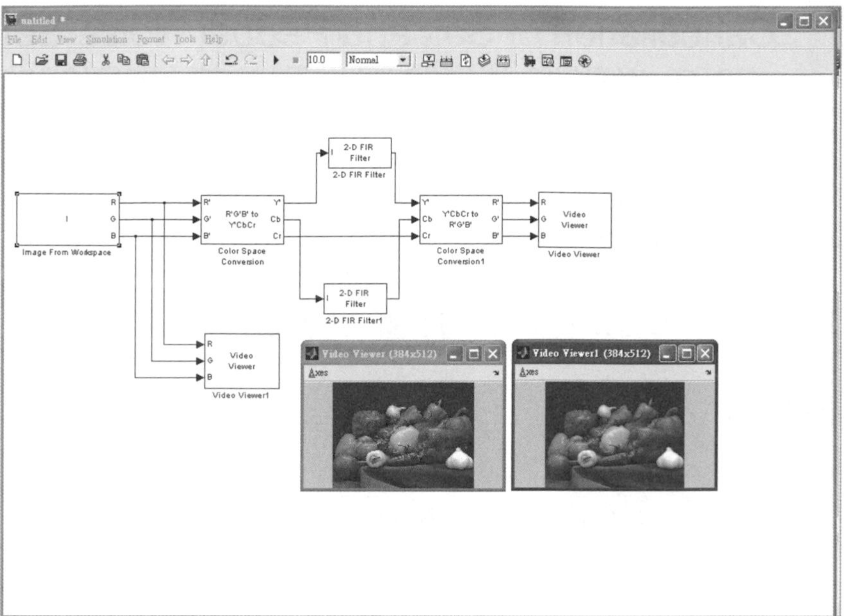
untitled *
File Edit View Simulation Format Tools Help
10.0
Normal
Image From Workspace
Color Space Conversion
R'G'B' to Y'CbCr
2-D FIR Filter
2-D FIR Filter1
Y'CbCr to R'G'B'
Color Space Conversion1
Video Viewer
Video Viewer1
Video Viewer (384x512)
Video Viewer1 (384x512)
Axes

6-2 Cancellation of Salt and Pepper Noise(Solution)【6-2 從影像中移除黑白點雜訊(問題解答)】

1. How can we make the output image perfect by adjusting the value of “Neighborhood size” in the “median filter” (2)?
【1. 如何藉由調整方塊(2) “median filter(均值濾波器)”的“Neighborhood size(相鄰尺寸)”值，使輸出影像趨於完美。】

- Hints(提示)

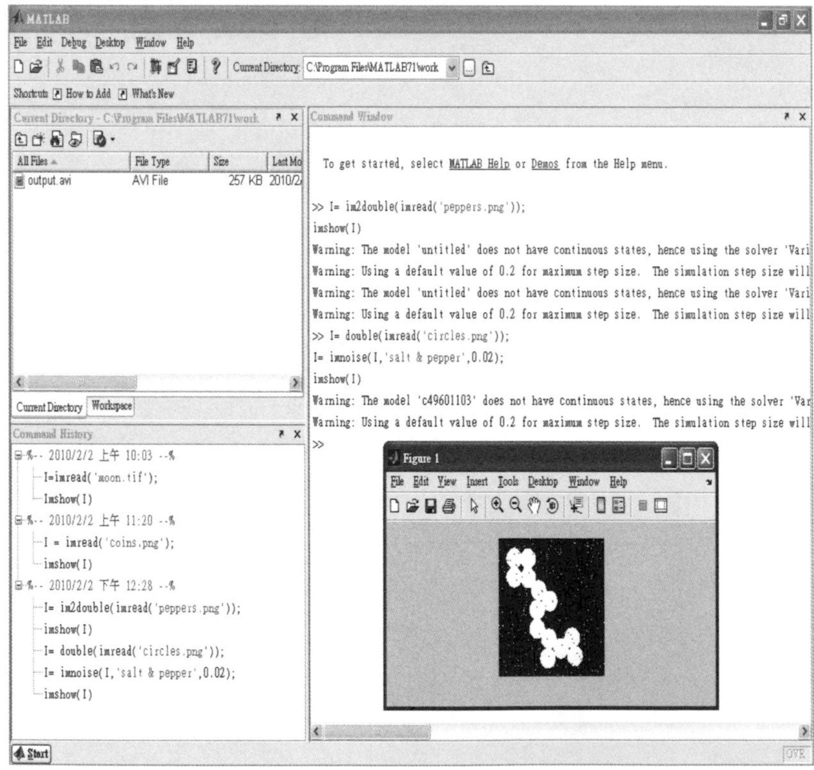

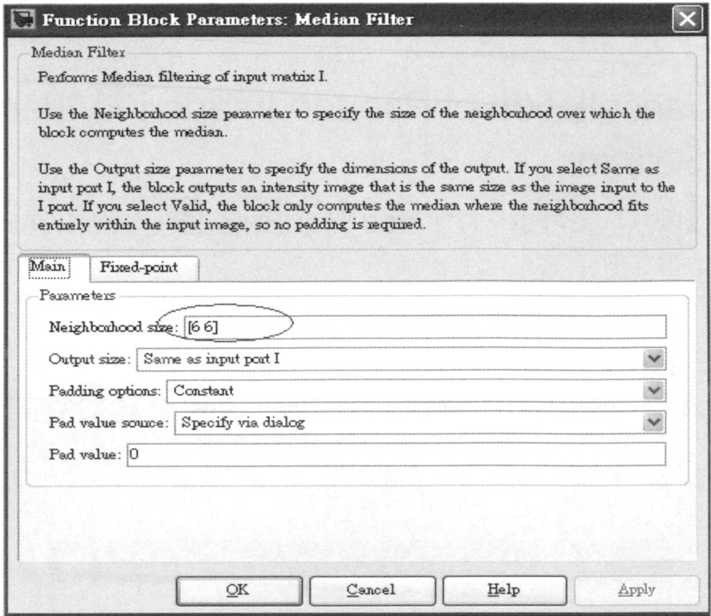
Function Block Parameters: Median Filter
Median Filter
Performs Median filtering of input matrix I.
Use the Neighborhood size parameter to specify the size of the neighborhood over which the block computes the median.
Use the Output size parameter to specify the dimensions of the output. If you select Same as input port I, the block outputs an intensity image that is the same size as the image input to the I port. If you select Valid, the block only computes the median where the neighborhood fits entirely within the input image, so no padding is required.
Main
Fixed-point
Parameters
Neighborhood size: [6 6]
Output size: Same as input port I
Padding options: Constant
Pad value source: Specify via dialog
Pad value: 0
OK
Cancel
Help
Apply

- Results(結果)

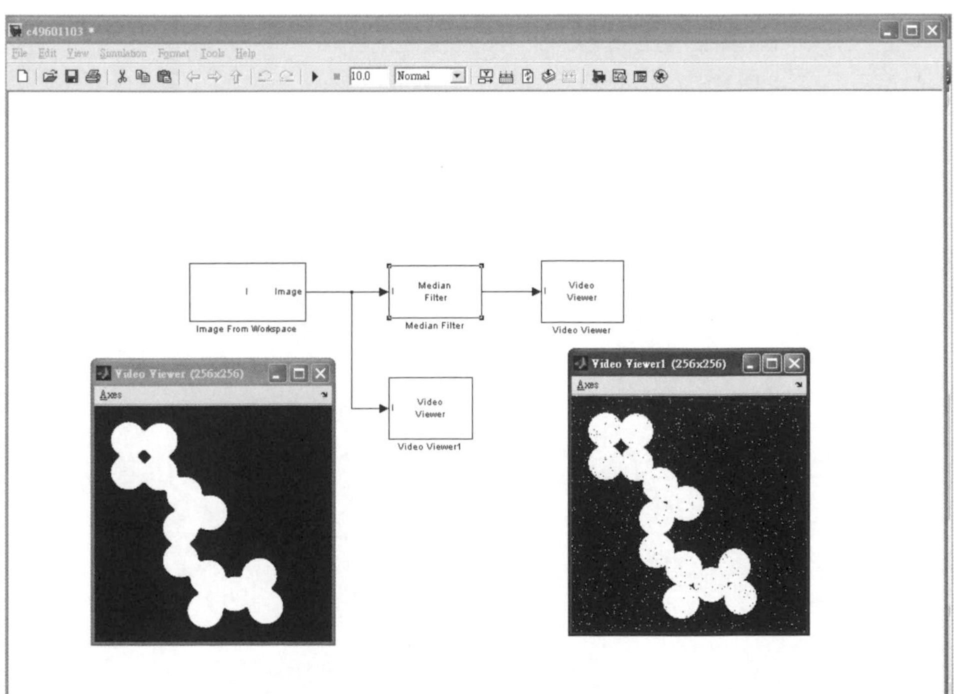
c49601103 *
File Edit View Simulation Format Tools Help
10.0
Normal
Image
Image From Workspace
Median Filter
Median Filter
Video Viewer
Video Viewer
Video Viewer
Video Viewer1
Video Viewer (256x256)
Video Viewer1 (256x256)
Axes

6-3 Cancellation of Pseudo Image in Video Signal (Solution)
【6-3 移除影片中的虛擬影像(問題解答)】

1. Please adjust the value of gain in block (9) to make the output image perfect.
   【1.請調整方塊(9)內的增益(Gain)值，使影像趨於完美。】

- Hints(提示)

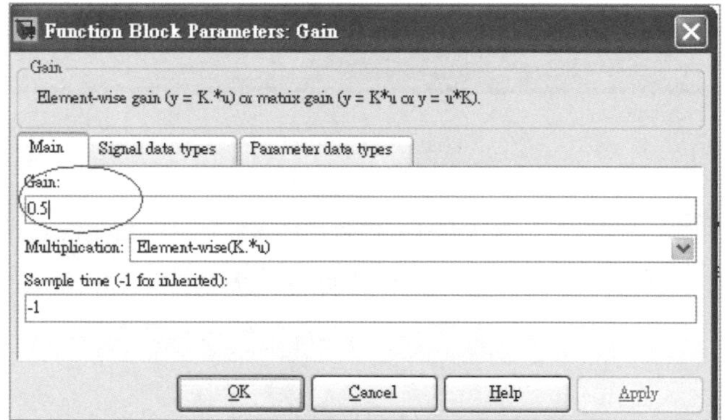

- Results(結果)

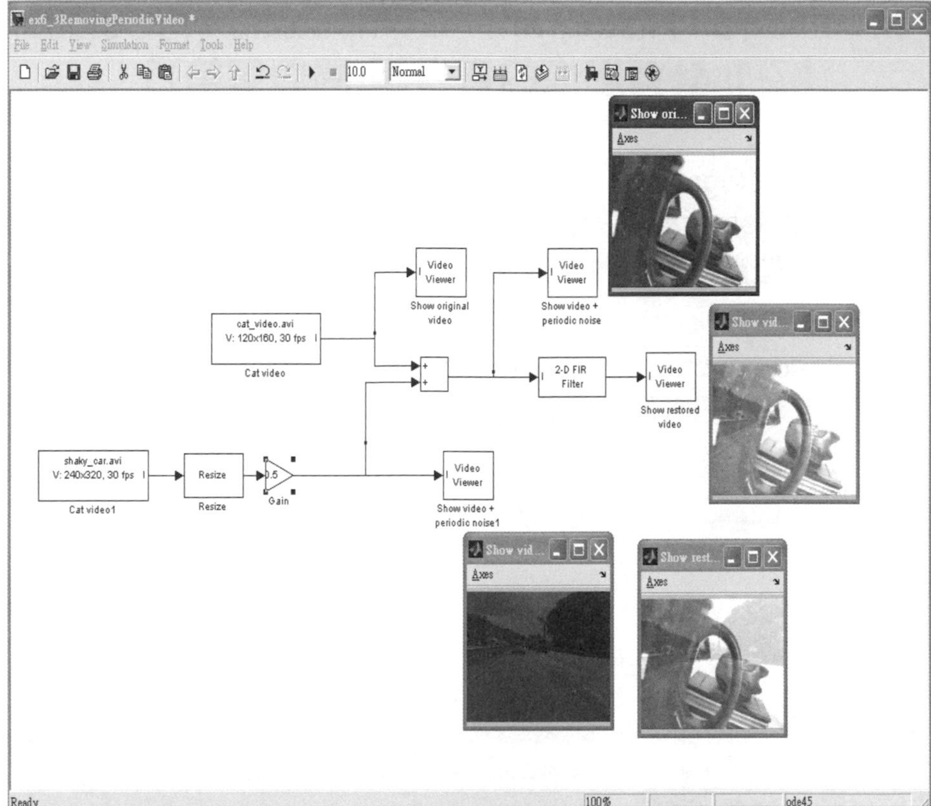
ex6_3RemovingPeriodicVideo *
File Edit View Simulation Format Tools Help
10.0
Normal
Show ori...
Axes
Video Viewer
Show original video
Video Viewer
Show video + periodic noise
cat_video.avi
V: 120x160, 30 fps
Cat video
2-D FIR Filter
Video Viewer
Show restored video
Show vid...
shaky_car.avi
V: 240x320, 30 fps
Cat video1
Resize
Resize
0.5
Gain
Video Viewer
Show video + periodic noise1
Show vid...
Show rest...
Ready
100%
ode45

7-1 Image Histogram Computation (Solution)
【7-1 計算影像的直方圖(問題解答)】

1. What is “histogram”?

   【1.何謂“histogram (直方圖)”？】將一組數據分成數組後，再繪製成連續型資料之次數分佈圖。

2. Download a picture, and compute its RGB histogram.

   【2.請下載一個圖片，重新計算其 RGB 直方圖。】

- Hints(提示)

```
>> I= im2double(imread('abc123.jpg'));

>> imshow(I)
```

- Results(結果)

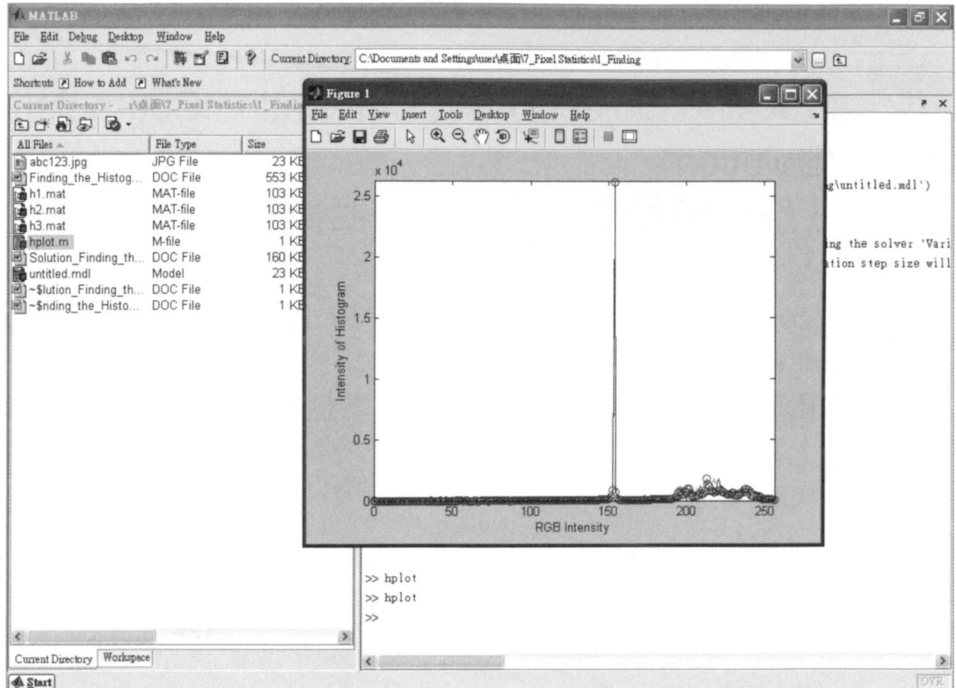
MATLAB
File Edit Debug Desktop Window Help
Current Directory: C:\Documents and Settings\user\桌面\7_Pixel Statistics\1_Finding
Figure 1
x 10^4
Intensity of Histogram
RGB Intensity
abc123.jpg
Finding_the_Histog...
h1.mat
h2.mat
h3.mat
hplot.m
Solution_Finding_th...
untitled.mdl
>> hplot
>> hplot
>>
Current Directory
Workspace

8-1 Tracking Object Using Correlation Computation (Solution)
【8-1 使用相關運算追蹤目標(問題解答)】

1. If “Constant value” is entered to be “[20 20]” in Constant block (6), what will be find in the output?
   【1.如果將方塊 6 “Constant value(常數值)=[20 20]”，輸出將有何變化？】

- Hints(提示)

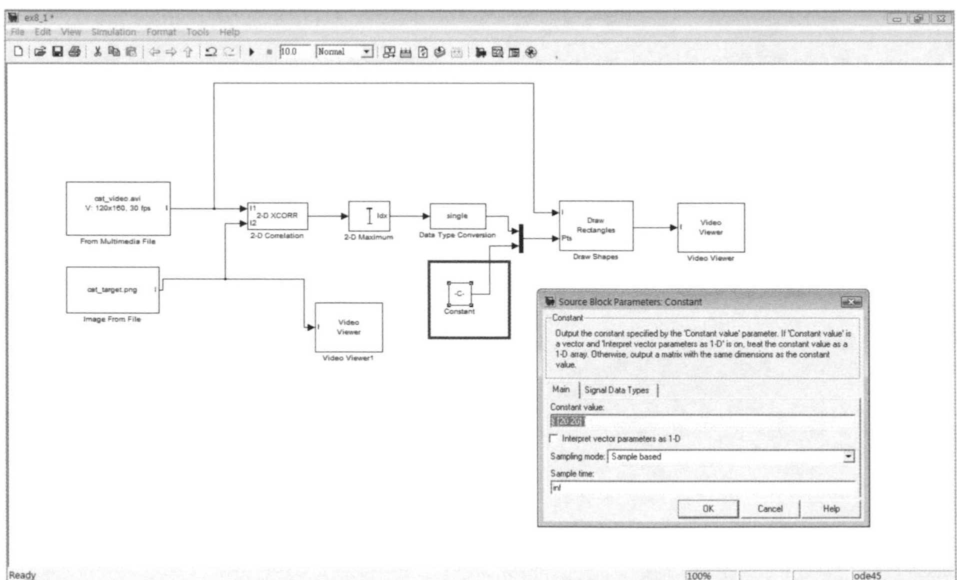

- Results(結果)

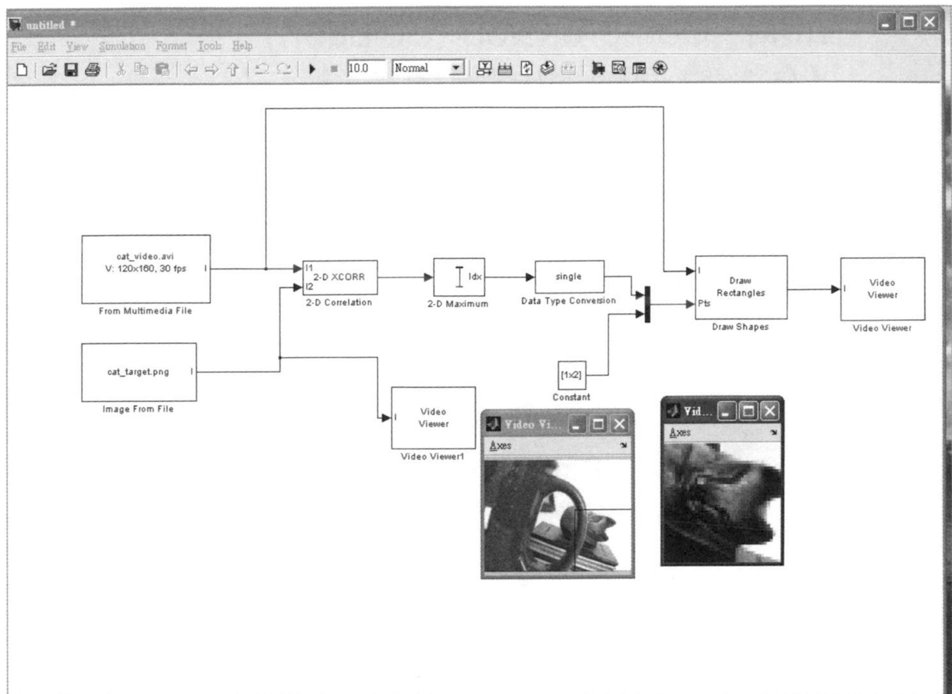
untitled *
File Edit View Simulation Format Tools Help
10.0
Normal
cat_video.avi
V: 120x160, 30 fps
From Multimedia File
2-D XCORR
2-D Correlation
Idx
2-D Maximum
single
Data Type Conversion
Draw
Rectangles
Pts
Draw Shapes
Video
Viewer
Video Viewer
cat_target.png
Image From File
[1x2]
Constant
Video
Viewer
Video Viewer1
Axes
Axes

9-1 Image Compression (Solution)【9-1 影像壓縮(問題解答)】

1. What will be changed in the output images if we set “overlap= {[4 4]}” in blocks (2) and (3)? Why?【1.將方塊 2 與方塊 3 中設定 “overlap(重疊範圍)= {[4 4]}”，則輸出將有何變化？】

- 1.Hints(提示)

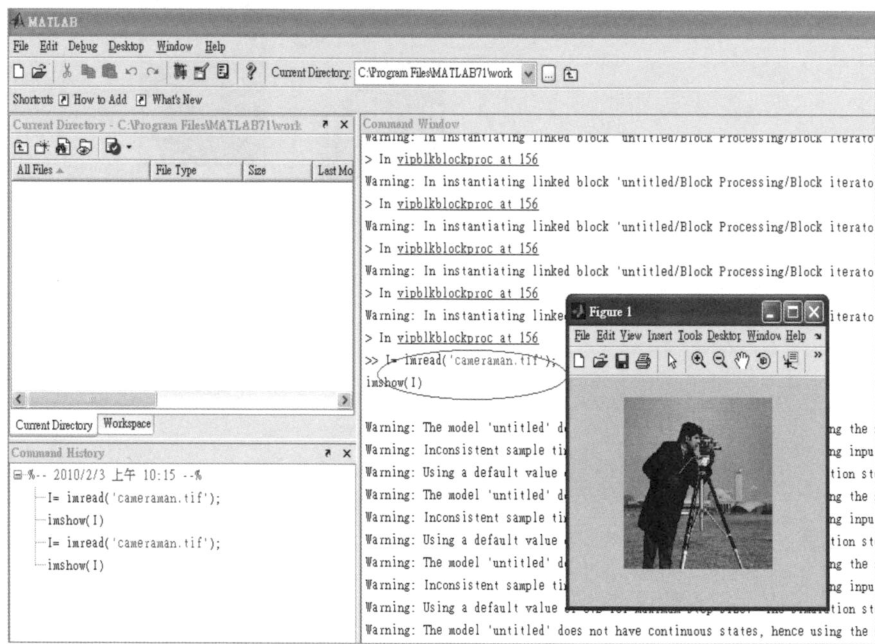

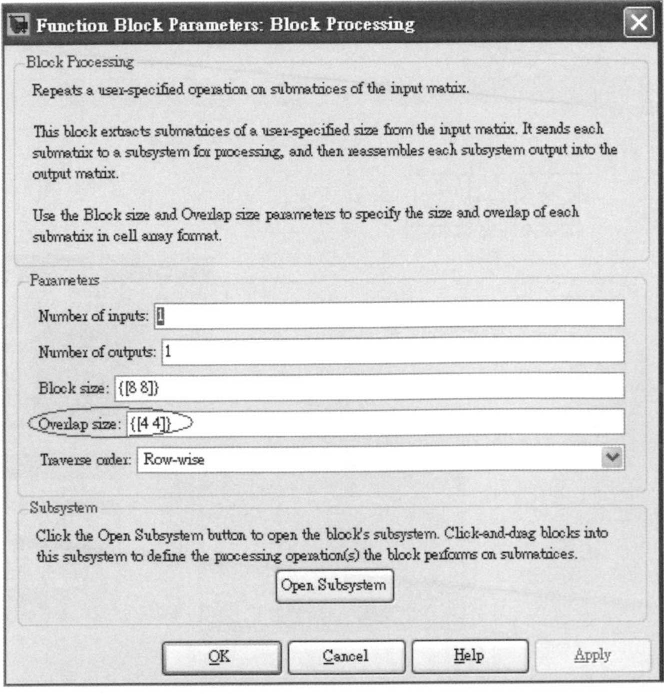
Function Block Parameters: Block Processing
Block Processing
Repeats a user-specified operation on submatrices of the input matrix.
This block extracts submatrices of a user-specified size from the input matrix. It sends each submatrix to a subsystem for processing, and then reassembles each subsystem output into the output matrix.
Use the Block size and Overlap size parameters to specify the size and overlap of each submatrix in cell array format.
Parameters
Number of inputs: 1
Number of outputs: 1
Block size: {[8 8]}
Overlap size: {[4 4]}
Traverse order: Row-wise
Subsystem
Click the Open Subsystem button to open the block's subsystem. Click-and-drag blocks into this subsystem to define the processing operation(s) the block performs on submatrices.
Open Subsystem
OK
Cancel
Help
Apply

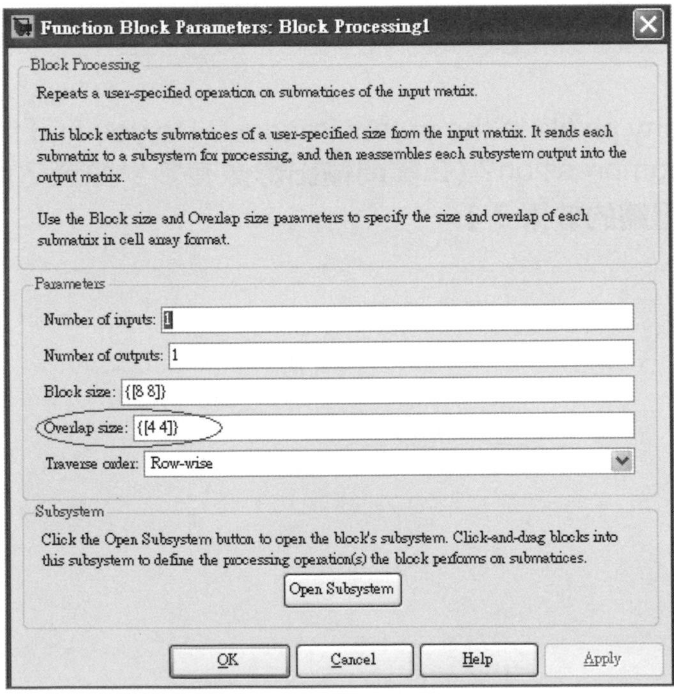
Function Block Parameters: Block Processing1
Block Processing
Repeats a user-specified operation on submatrices of the input matrix.
This block extracts submatrices of a user-specified size from the input matrix. It sends each submatrix to a subsystem for processing, and then reassembles each subsystem output into the output matrix.
Use the Block size and Overlap size parameters to specify the size and overlap of each submatrix in cell array format.
Parameters
Number of inputs: 1
Number of outputs: 1
Block size: {[8 8]}
Overlap size: {[4 4]}
Traverse order: Row-wise
Subsystem
Click the Open Subsystem button to open the block's subsystem. Click-and-drag blocks into this subsystem to define the processing operation(s) the block performs on submatrices.
Open Subsystem
OK
Cancel
Help
Apply

- 1. Result(結果)

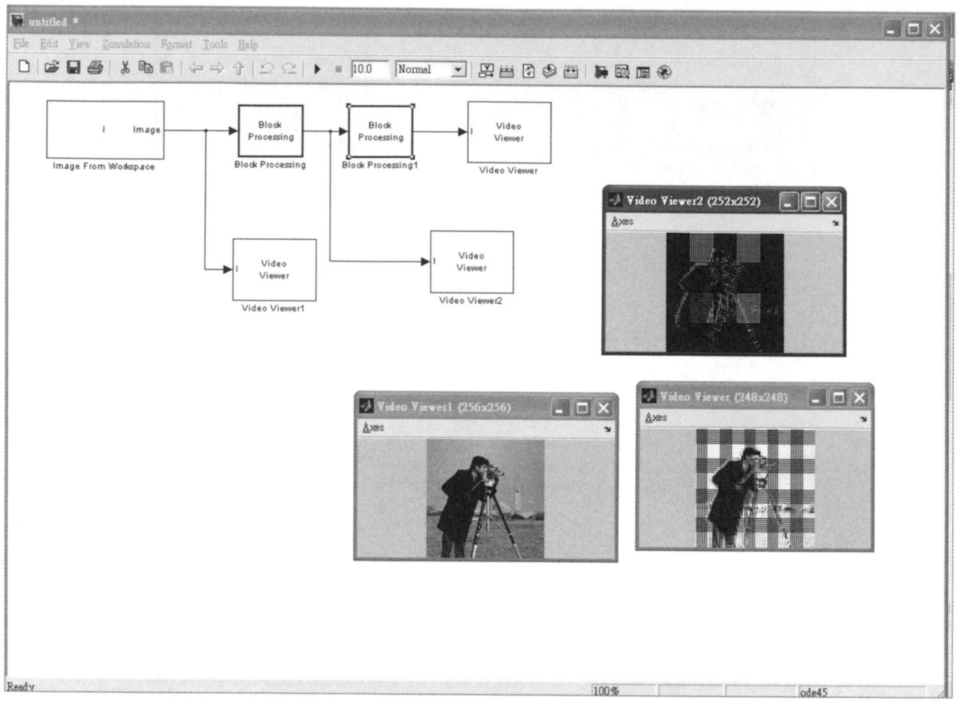

2. How many pixels in the output images? Do you find the effect of image compression?【2.請問輸出影像有多少像素？你是否有發現影像壓縮的效果？】

- 2.Result(結果)

(1)

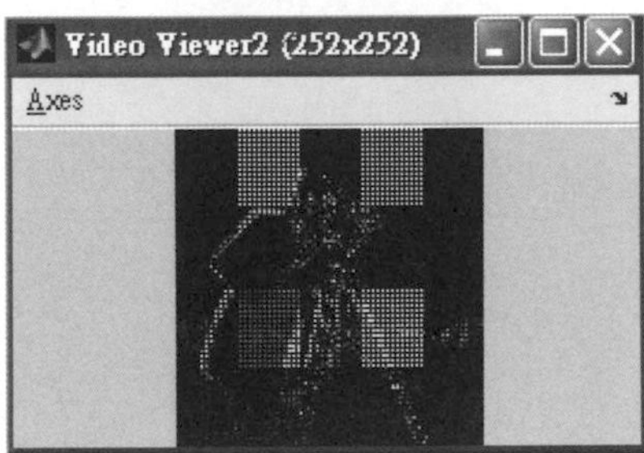

(2) Result(結果):當然會壓縮圖像

Amplitude / Frequency / Phase Modulation (Solution)
【10-1 振幅/頻率/相位 調變(問題解答)】

1. If the “Frequency deviation=3” in the block (8), what will be changed in the output signal of FM? Why?【1.如果將方塊 8 中設定“Frequency deviation(頻率偏移)=3”，FM 的輸出將有何變化？】

- Hints(提示)

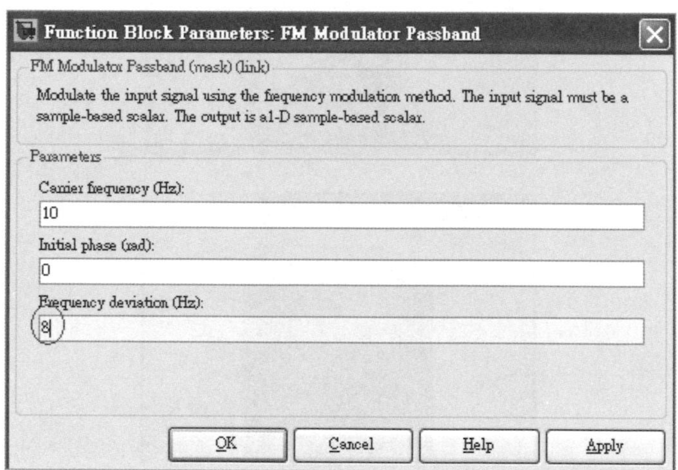

- Results(結果)

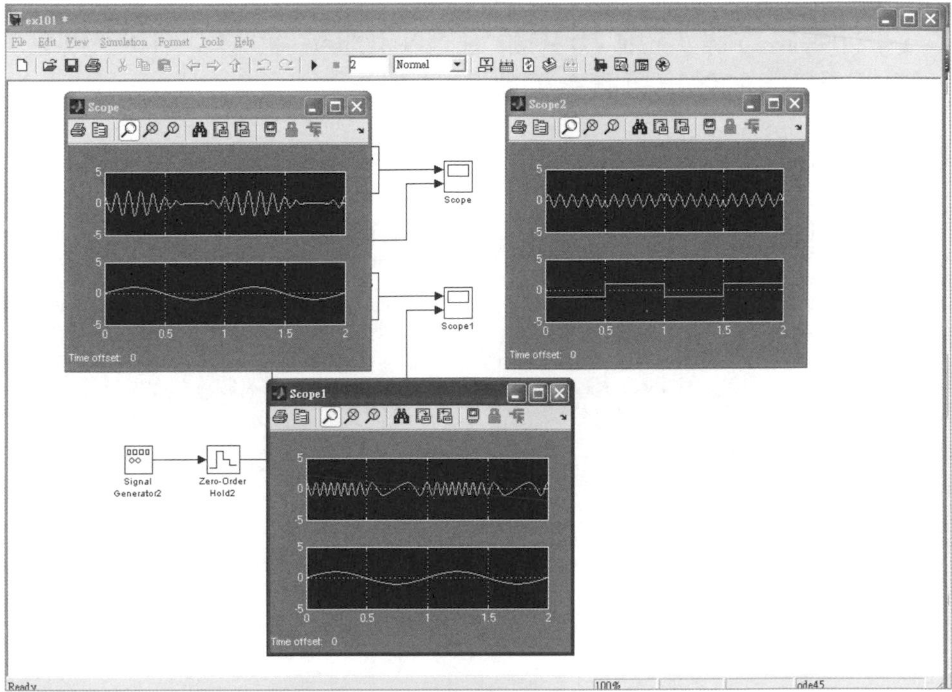

10-2 Amplitude/Frequency/Phase Modulation and Demodulation with Additive White Gaussian Noise (AM/FM/PM with AWGN) (Solution)【10-2 加入式白高斯雜訊混入通道中的振幅/頻率/相位調變與解調(問題解答)】

1. What is the output signals in blocks (19) to (21), if the "SNR=10" in the blocks (13) to (15)?
   【1.如果將方塊 13 到 15 中設定"SNR(訊號雜訊比)=10"，則在方塊 19 到 21 的輸出訊號有何變化？】

- Hints(提示)

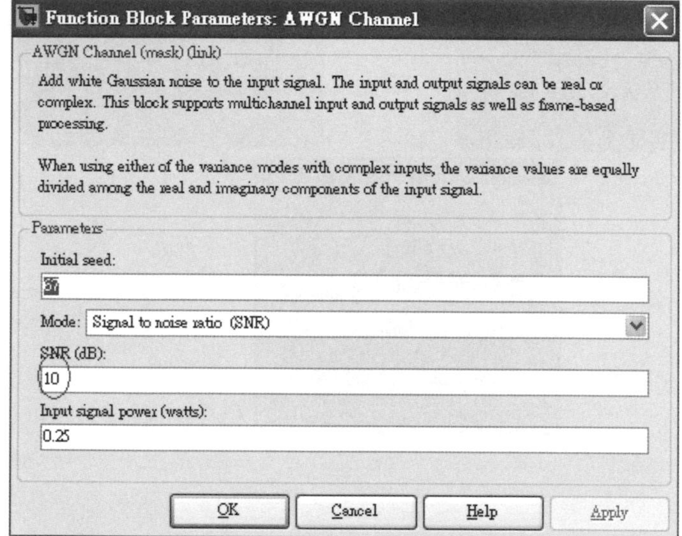

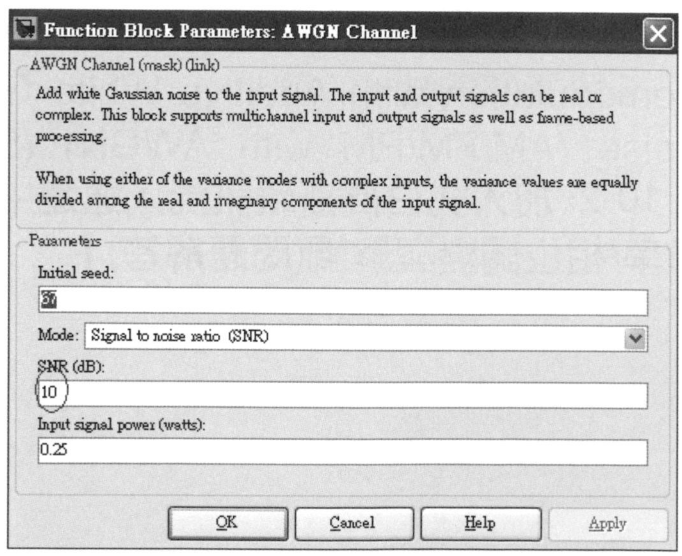

- Results(結果)

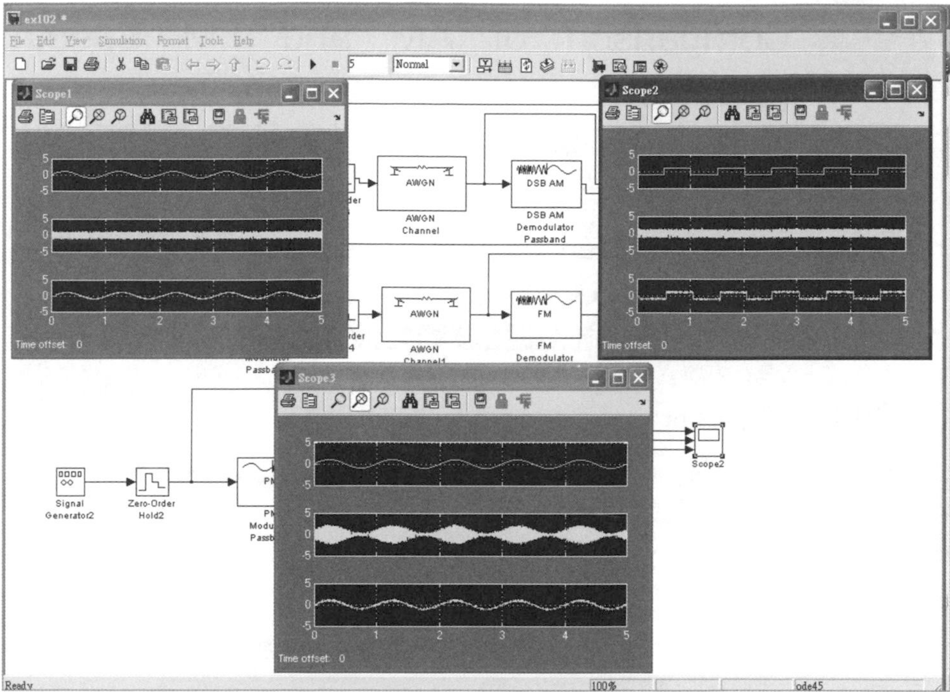
ex102 *
File Edit View Simulation Format Tools Help
Normal
Scope1
Time offset: 0
Scope2
Time offset: 0
Scope3
Time offset: 0
AWGN
AWGN Channel
DSB AM
DSB AM Demodulator Passband
AWGN
AWGN Channel1
FM
FM Demodulator
Signal Generator2
Zero-Order Hold2
Scope2
Ready
100%
ode45

10-3 Comparison of AM and FM Spectrums (Solution)

【10-3 AM 與 FM 的頻譜之比較(問題解答)】

1. What are the differences of the spectrums of sine and square waves from Signal Generator?
   【1.從訊號產生器輸出正弦波與方波,其頻譜有何不同？】

- 1.Hints(提示) sine

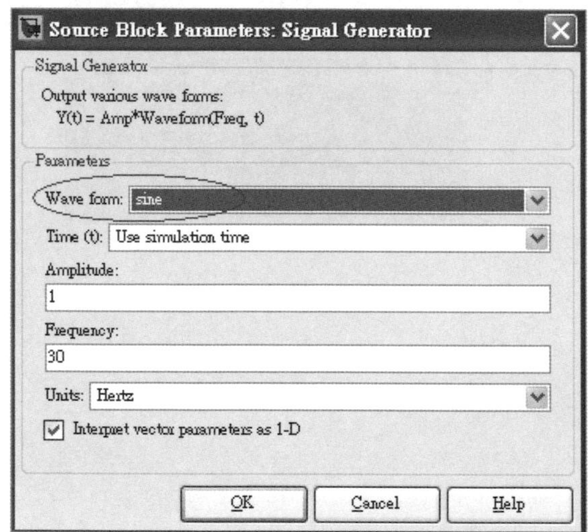

- Results(結果)

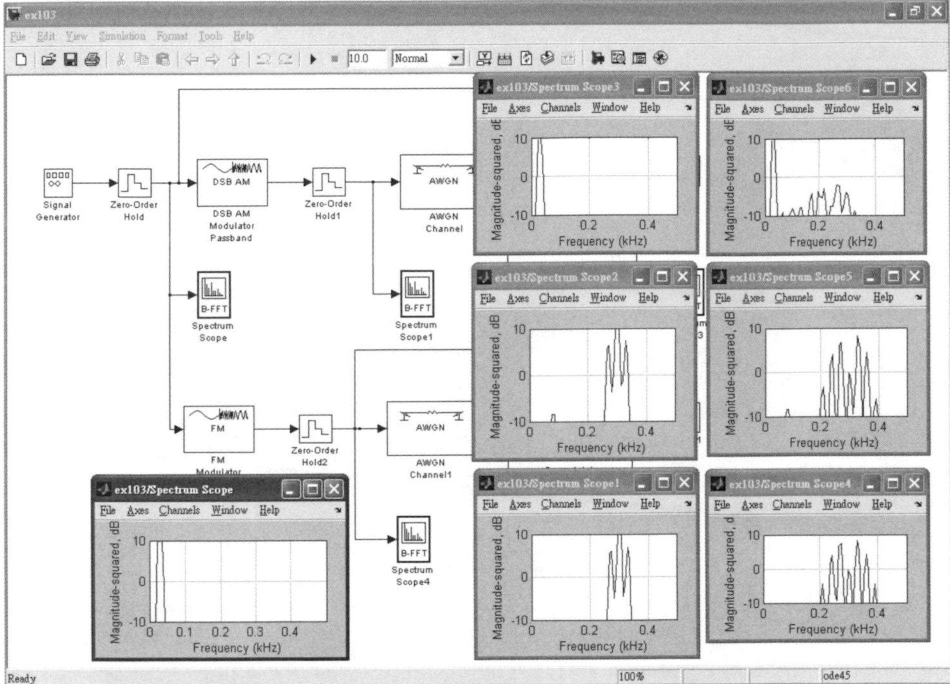

- Hints(提示) square

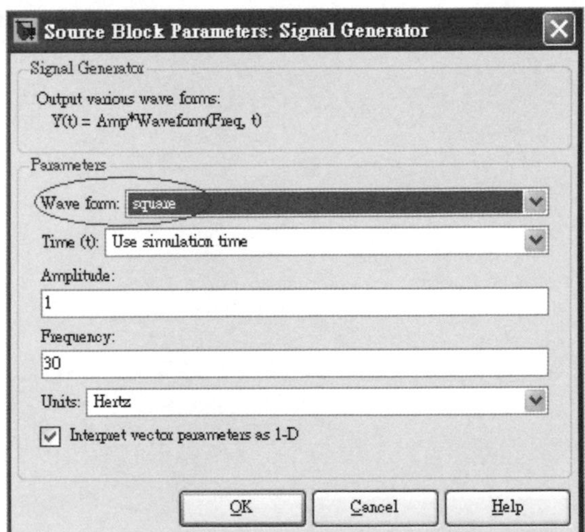

- Results(結果)

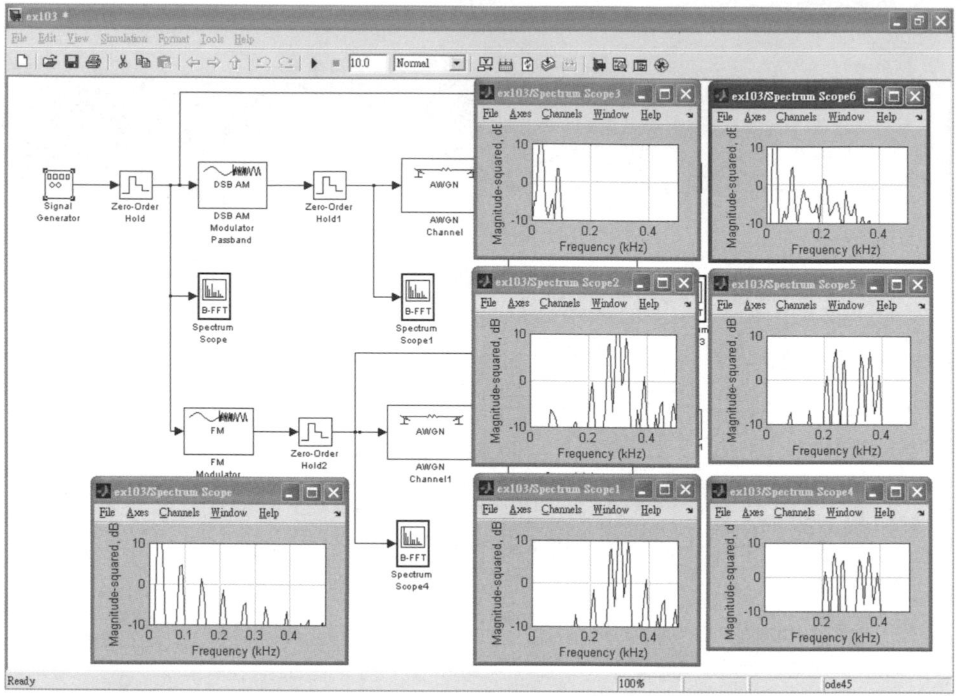

2. Compare the results of SNR=10, and 30 in AWGN channels.
   【2.比較在 AWGN 通道的 SNR(訊號雜訊比)=10 與 30,其輸出結果有何不同？】

- 2. Hints(提示) SNR=10 上下的 AWGN

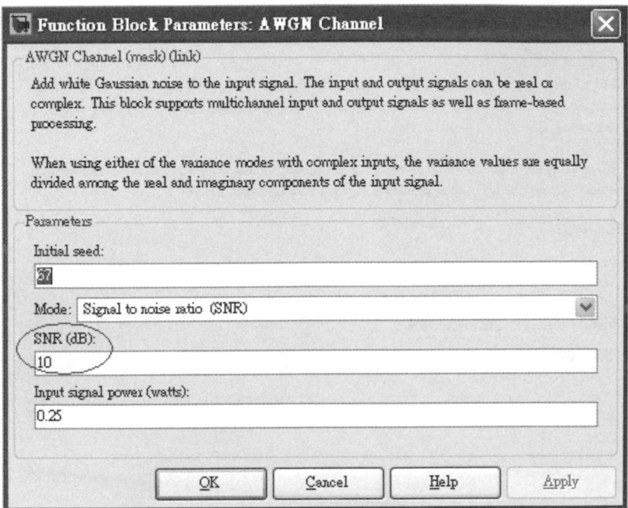

- Results(結果) SNR=10

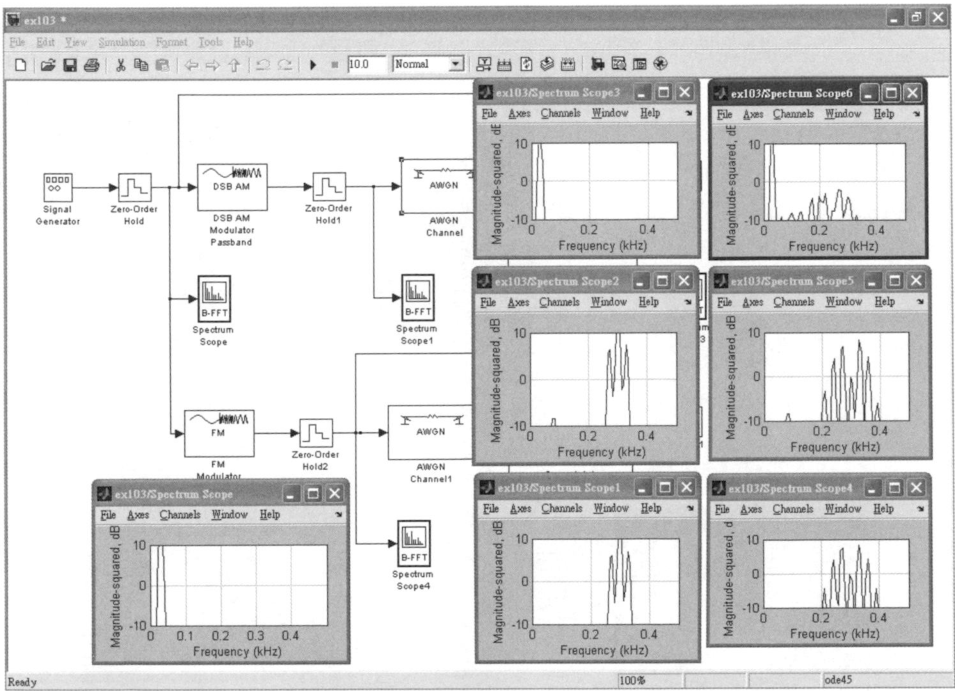

- Hints(提示) SNR=30 上下的 AWGN

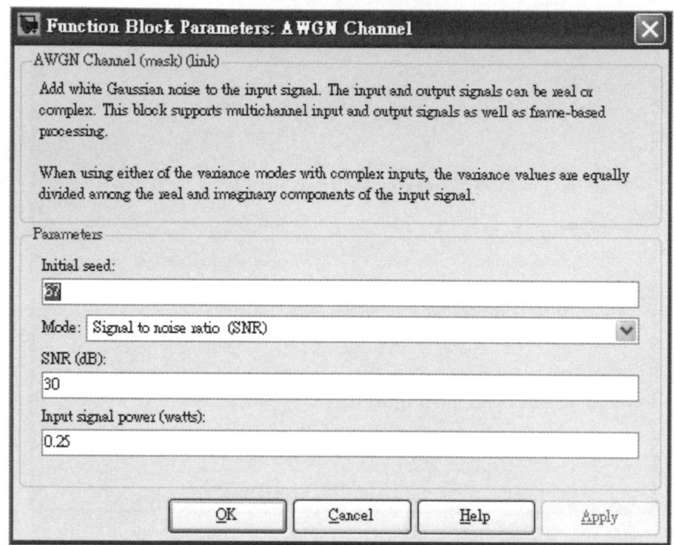

- Results(結果) SNR=30

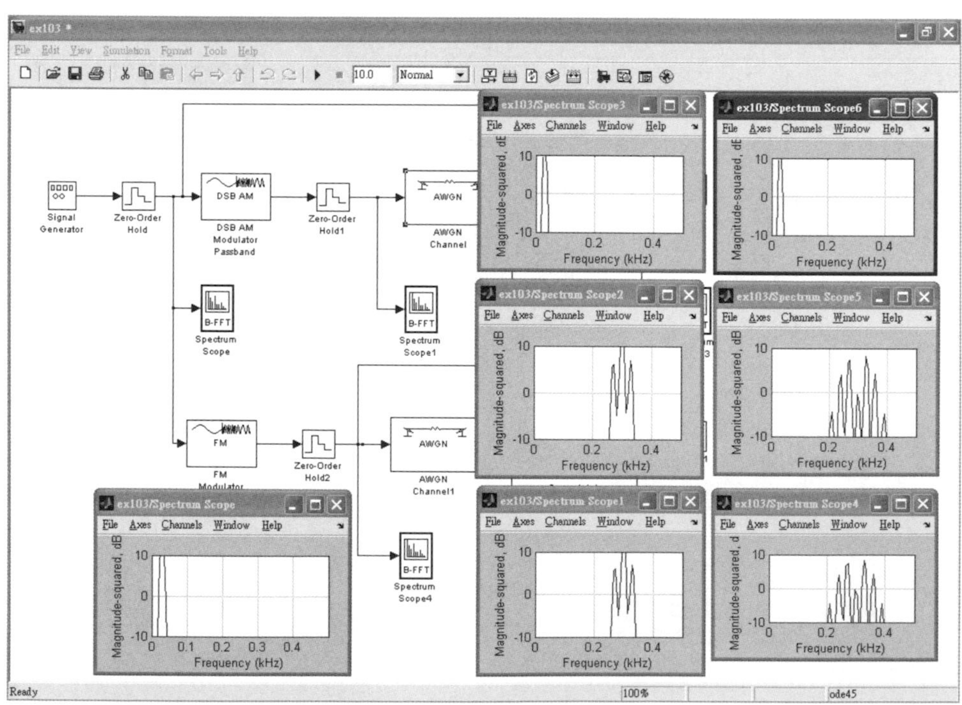

3. What are the differences of the filter order=4, and 10 in the results of AM Demodulations?
【3.在 AM 解調中，filter order(濾波器級數)=4 與 10 其輸出結果有何不同？】

- 3. Hints(提示) filter order=4

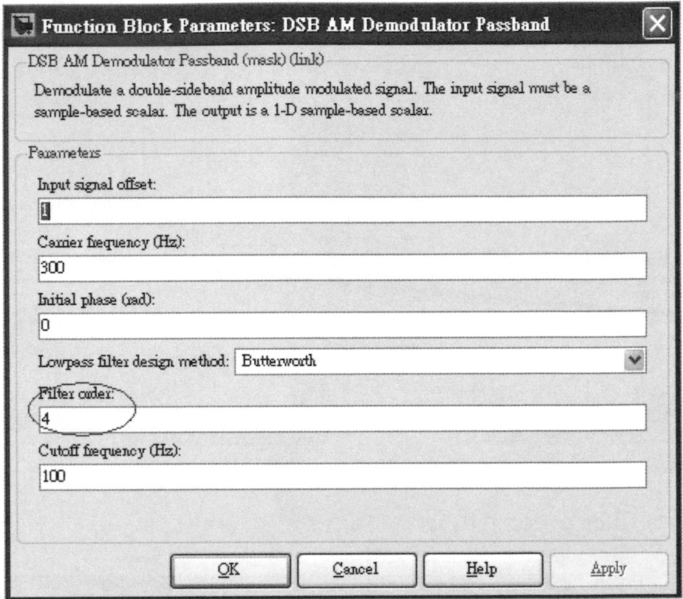

- Results(結果) filter order=4

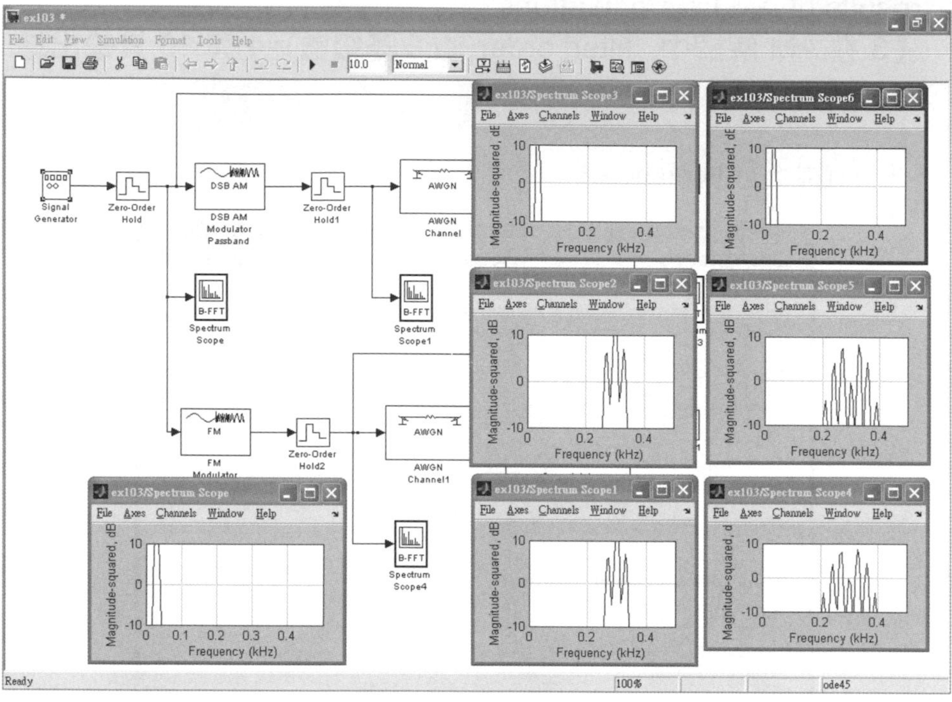

- Hints(提示) filter order=10

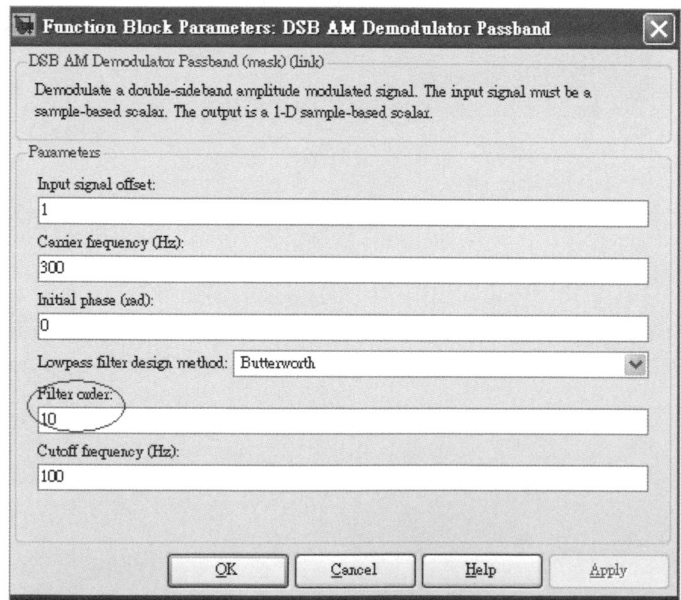

- Results(結果) filter order=10

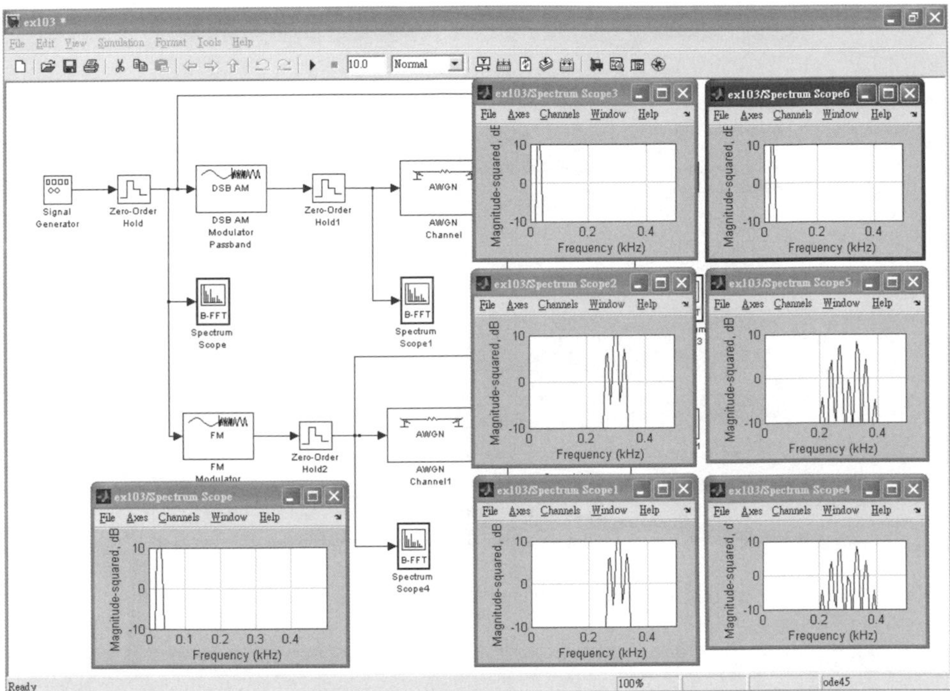

4. What are the differences of output signals of input signal power=0.25, and 0.05 from AWGN channels?
【4.比較在 AWGN 通道的訊號功率=0.25 與 0.05,其輸出結果有何不同？】

- 4. Hints(提示) signal power=0.25 上下都要改

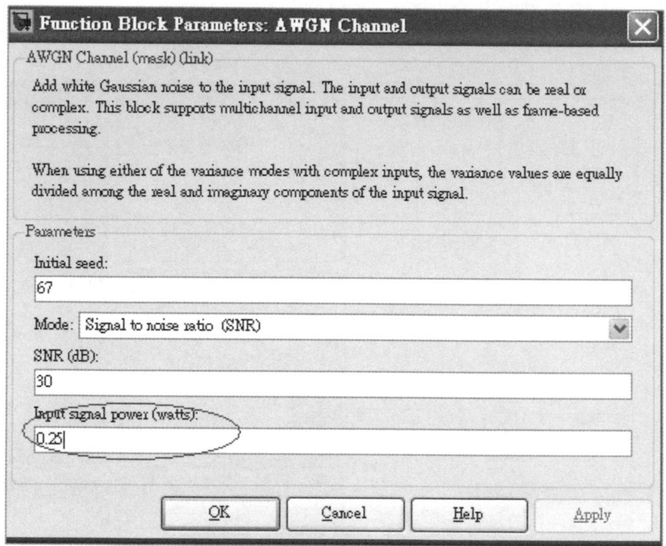

- Results(結果) signal power=0.25

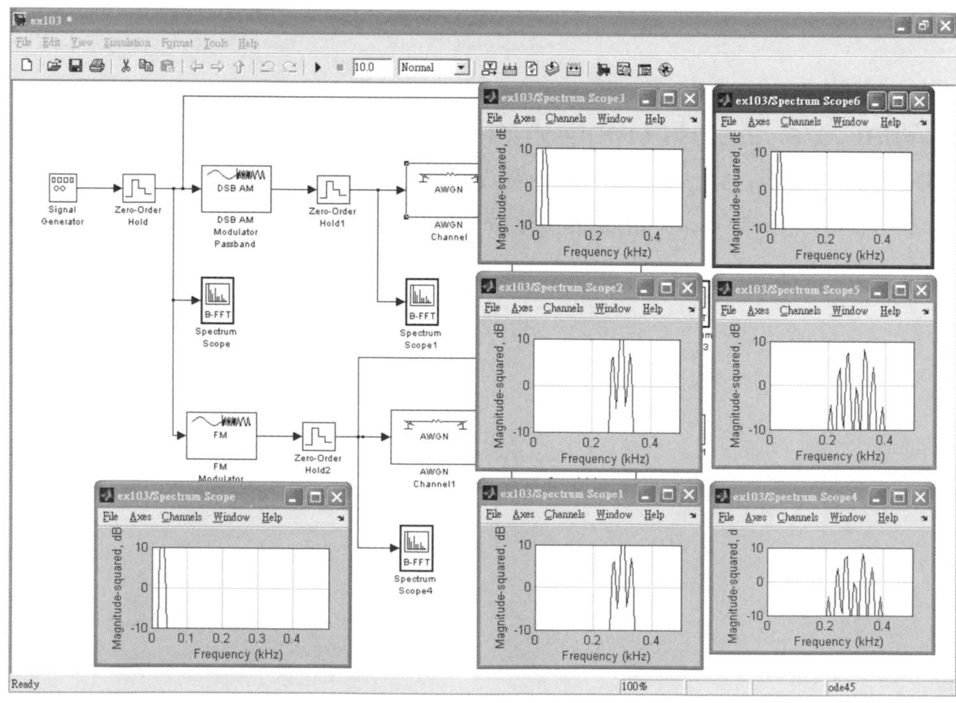

- Hints(提示) signal power=0.05

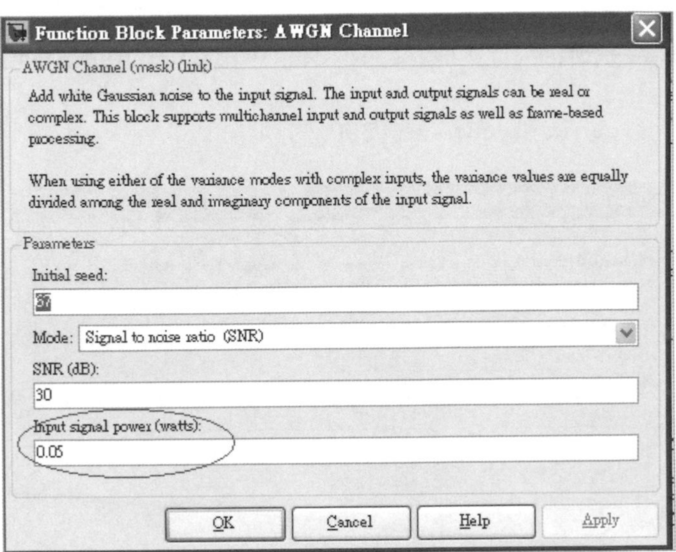

- Results(結果) signal power=0.05

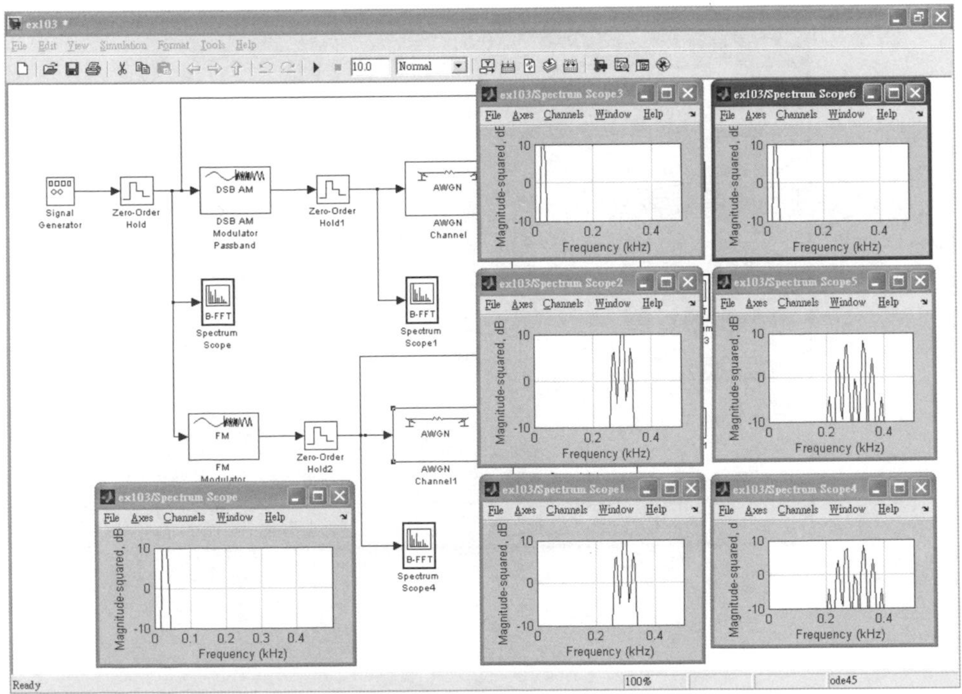

5. What are the differences of carrier frequencies=300 and 500 Hz in the results of AM, and FM demodulations?
【5.如果載波頻率為300與500,則AM與FM的輸出有何差異？】

- 5.Hints(提示) carrier frequencies=300

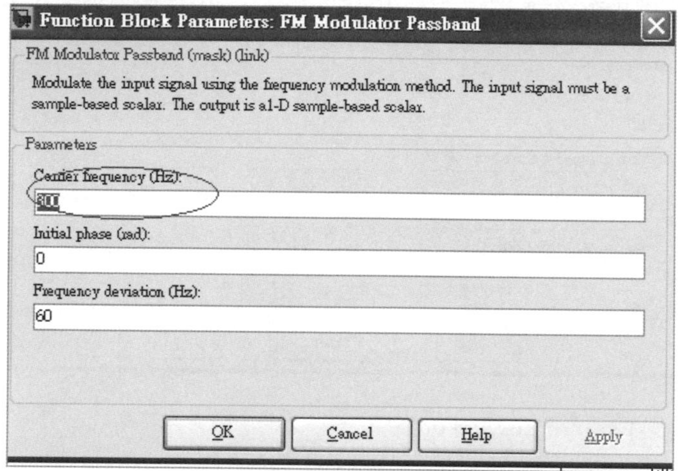

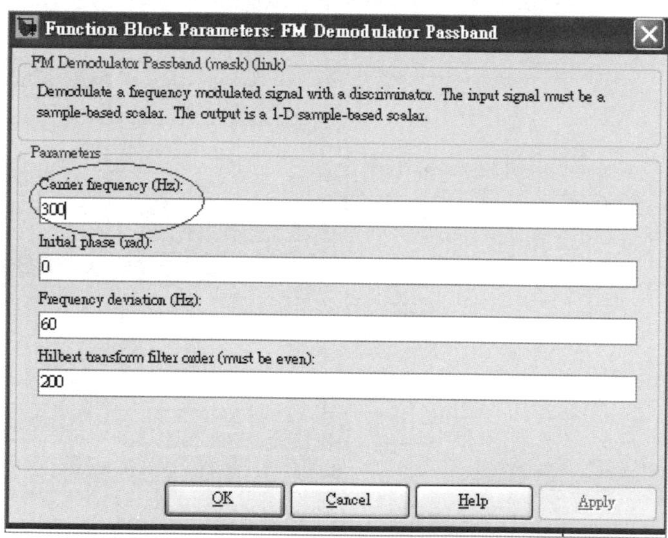

- Results(結果) carrier frequencies=300

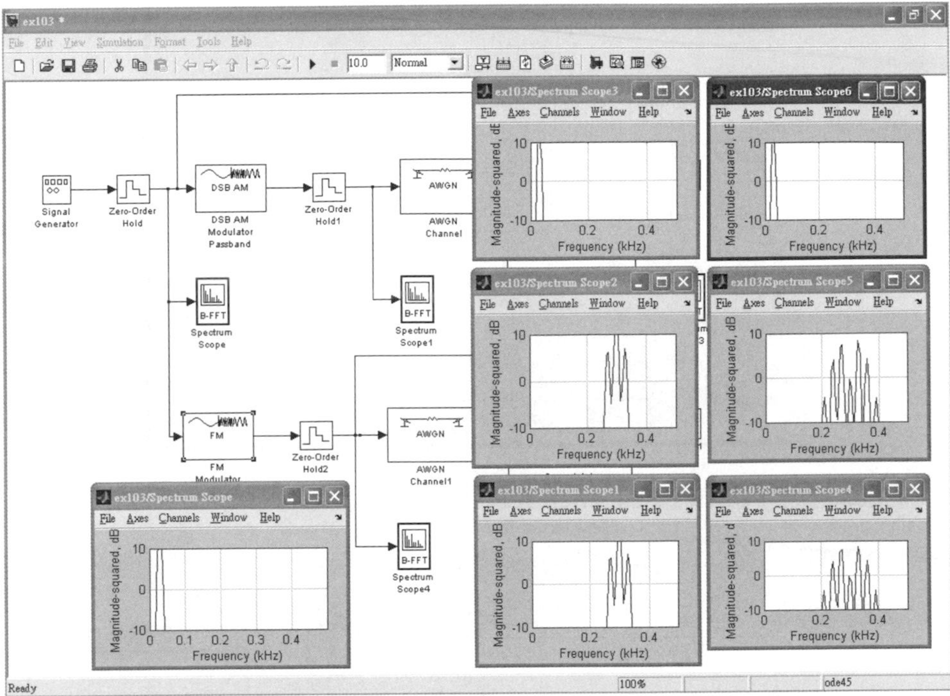

- Hints(提示) carrier frequencies=500

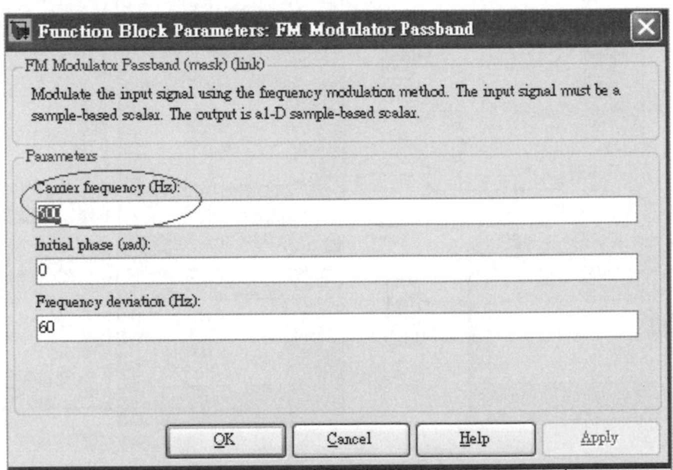

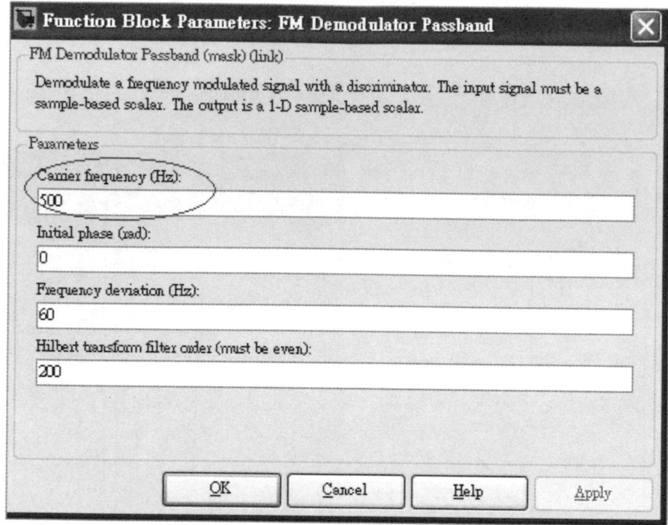

- Results(結果) carrier frequencies=500

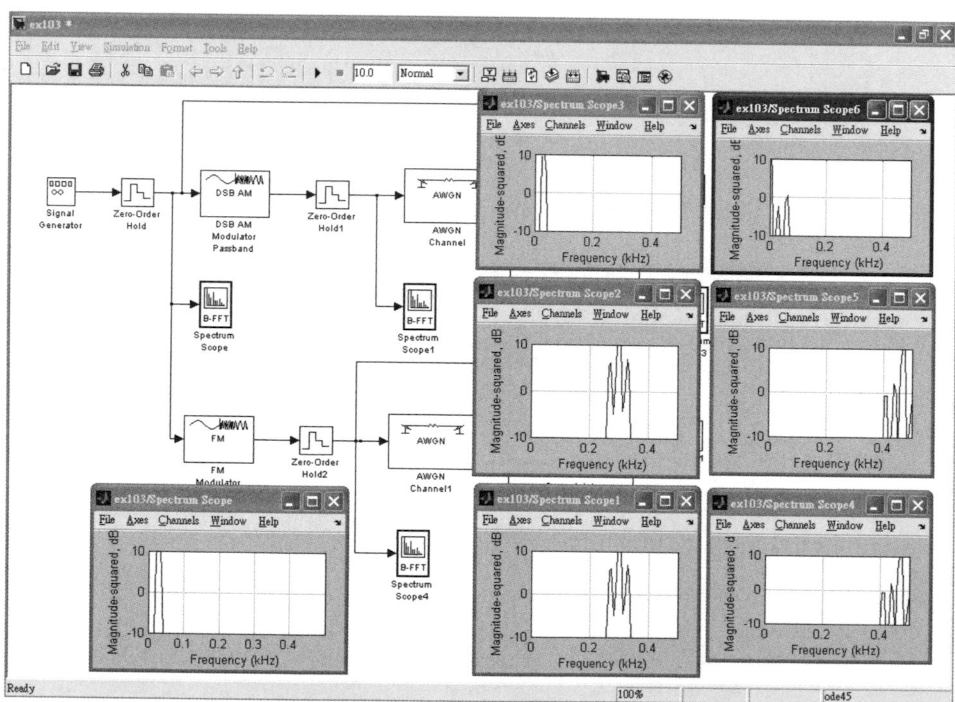

10-4 Double-Sidedband AM and Double-Sidedband Suppressed-Carrier AM (DSB AM & DSBSC AM)(Solution)【10-4 雙旁帶振幅調變與雙旁帶抑制載波振幅調變(問題解答)】

1. What are the differences of the spectrums of sine, square, and sawtooth waves from Signal Generator?【1.如果訊號產生器輸出正弦波、方波與鋸齒波，則輸出的頻譜有何差異？】

- 1.Hint(提示)sine

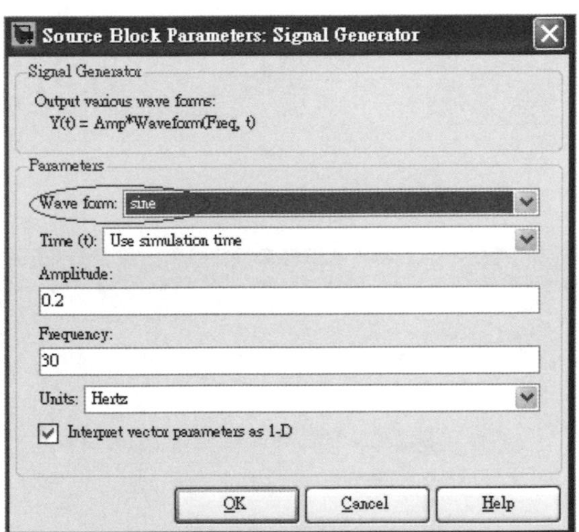

- Results(結果)sine

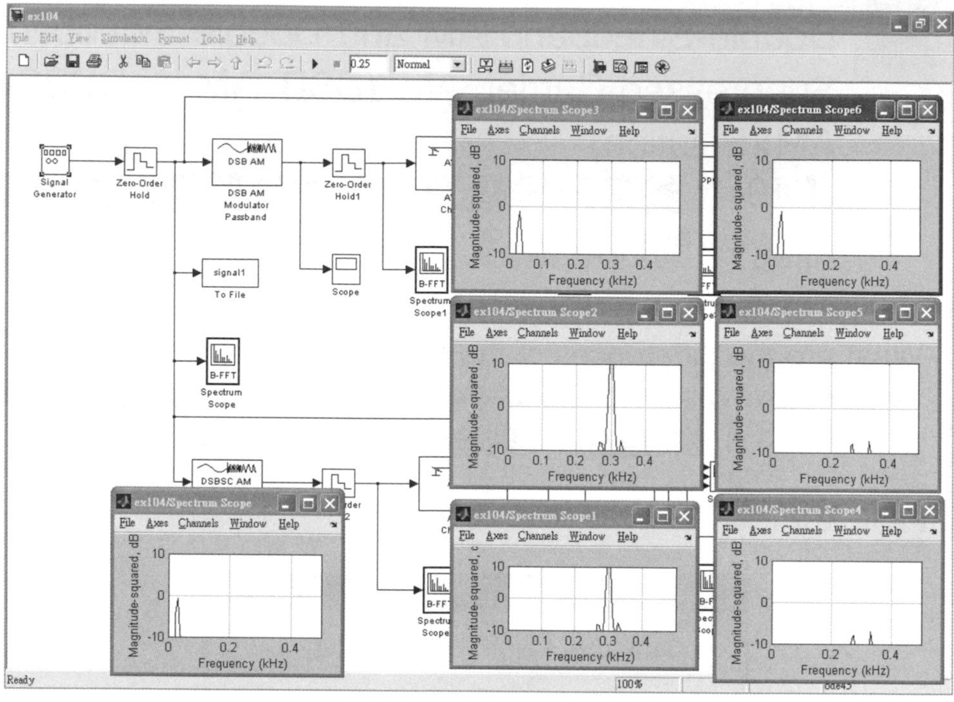

- Hint(提示) square

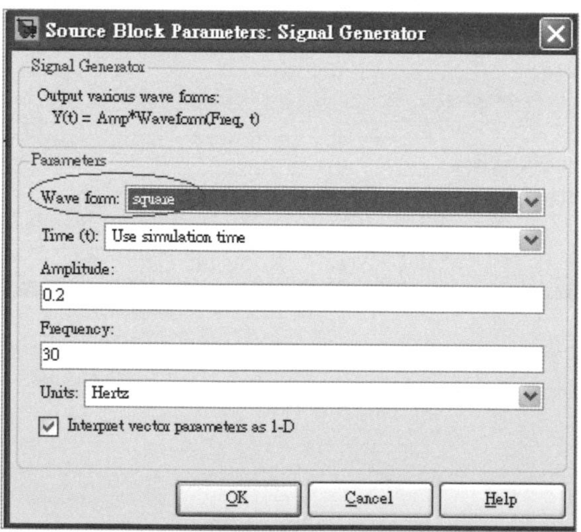

- Results(結果) square

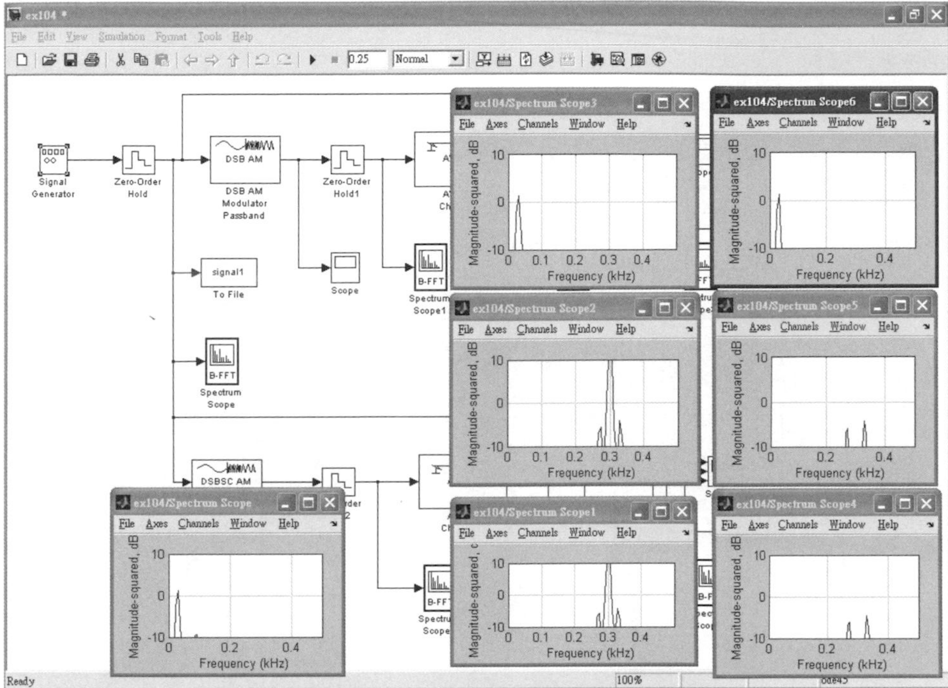

- Hint(提示) sawtooth

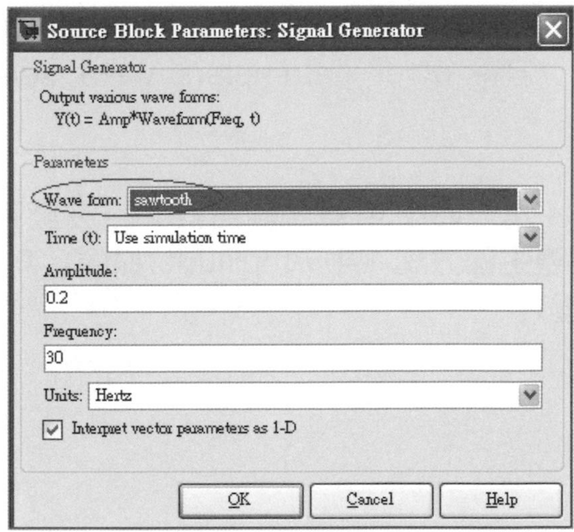
Source Block Parameters: Signal Generator

Signal Generator

Output various wave forms:
Y(t) = Amp*Waveform(Freq, t)

Parameters

Wave form: sawtooth

Time (t): Use simulation time

Amplitude:
0.2

Frequency:
30

Units: Hertz

☑ Interpret vector parameters as 1-D

OK | Cancel | Help

- Results(結果) sawtooth

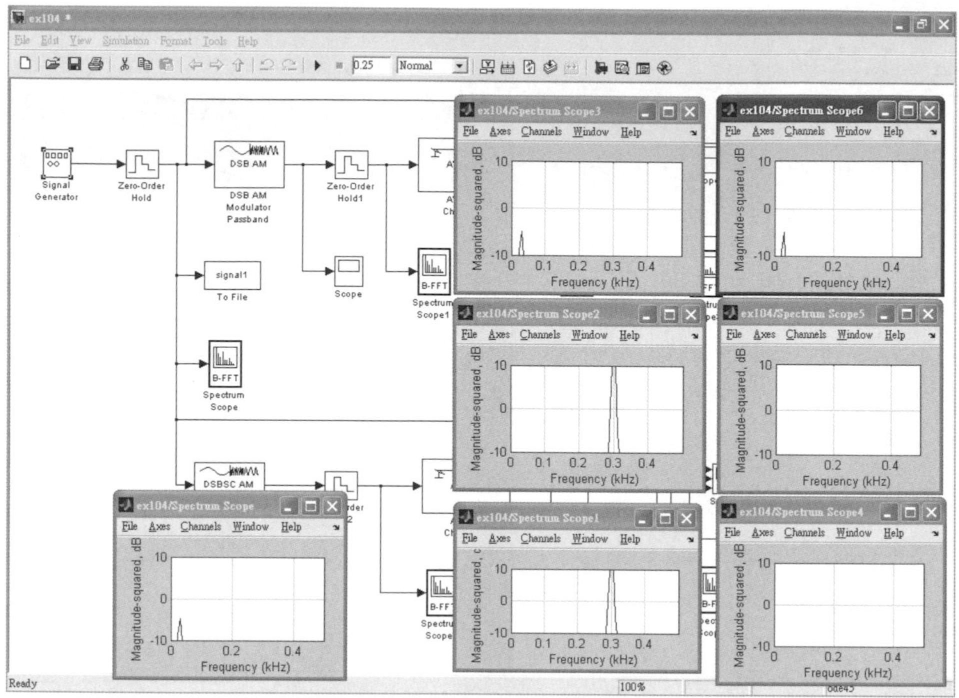

2. Compute the modulation index (m) form scope (14). Amplitude= 0.2, 0.5, and 1 in signal generator (1) in the DSB AM system. (The other parameters were selected the first ones.)
   where
   【2.從方塊 14 示波器中,計算調變指數(m)。在雙旁帶振幅調變的系統中，訊號產生器方塊 1 中 Amplitude(振幅)= 0.2, 0.5,與 1 (其他的參數不變)】

- 2.Results(結果) Amplitude= 0.2

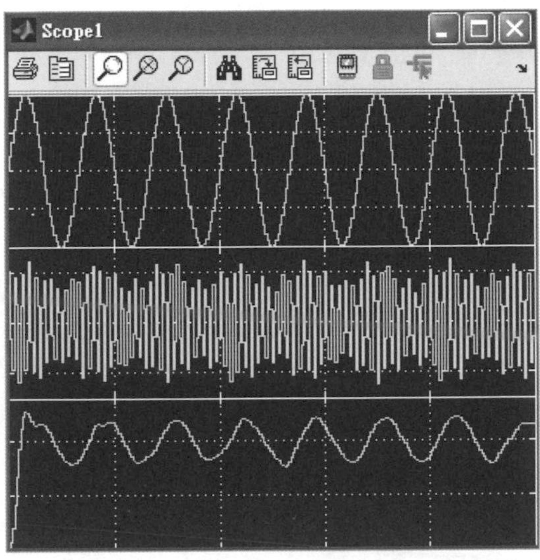

Max=2.2 Min=1.3 Ans=0.26

- Results(結果) Amplitude= 0.5

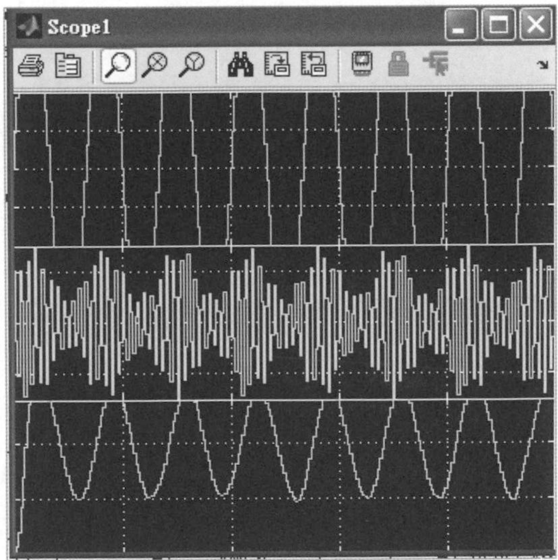

Max=3 Min=1. Ans=0.5

3. Find the main differences of DSB AM and DSBSC AM.
   【3.請指出 DSB AM(雙旁帶振幅調變)與 DSBSC AM(抑制載波雙旁帶振幅調變)有何差異？】

- 3.Results(結果)Scope2 跟 5 不同

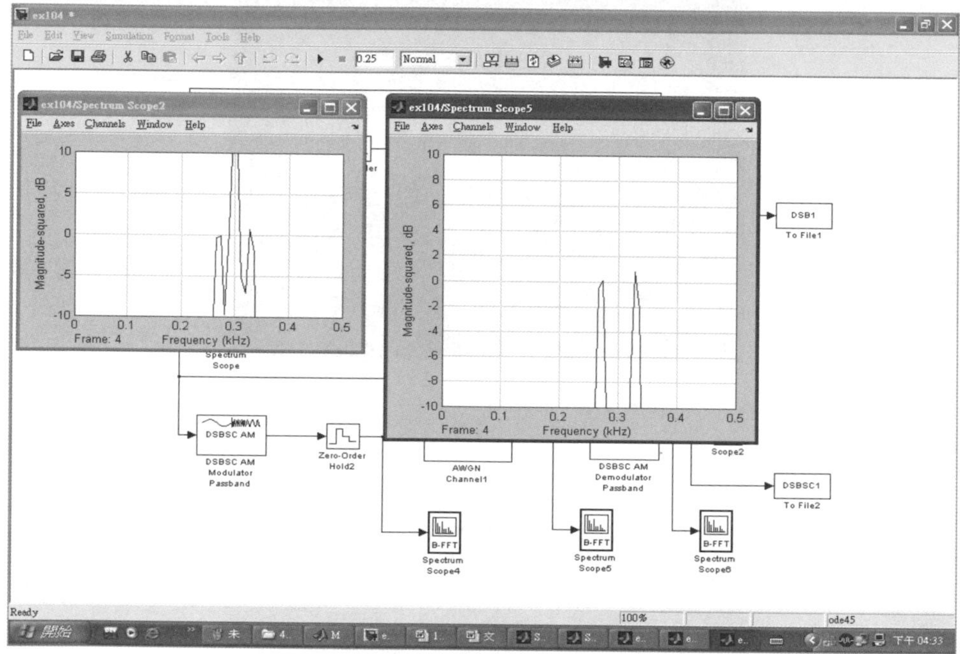

4. What do you find in the 2 figures from the results of the following program?
   【4.請執行以下的程式尋找圖形結果的差異。】

- 4.Results(結果)

Matlab→file→New→M-file (Editor untitled)

```
% begin the program
% Programer: Benjamin Bing Yuh Lu
clear all;
load signal1
load DSB1
```

```
load DSBSC1
figure(1)
subplot(311)
plot(s1(1,:), s1(2,:))
subplot(312)
plot(DSB1(1,:), DSB1(2,:))
subplot(313)
plot(DSBSC1(1,:), DSBSC1(2,:))
figure(2)
subplot(211)
plot(s1(1,:),s1(2,:)-DSB1(2,:))
subplot(212)
plot(s1(1,:),s1(2,:)-DSBSC1(2,:))
% end the program
```

筆記頁

Vocabulary
字彙

1-1
Default 預設值
View 觀看
Link 連結
Video 影像
Enhancement 加強
Image 影像
Process 處理
File 檔案
Strengthen 強化
Connect 連接

1-2
Parameter 參數
Show 顯示
Create 創造
Enable 能
Model 模型
Section 組,分組
Value 價值
Set 設定
Double 雙倍的,雙倍精度數
Play 放映,執行,播放

1-3
Multimedia 多媒體
Standard 標準
Audio 聲頻
Viewer 觀看者
Illustration 説明
Inverter 反向器
Happen 發生
Verify 查證
Keep 保持
Surround 包圍

1-4
Export 輸出
Figure 圖
Procedure 程序
Desktop 桌面
Assume 假設
Equivalent 等效的
Already 已經
Configuration 組態
Setting 設定
Directory 目錄

2-1
Present 顯示,展現
Watch 看,注意
Image From Workspace 從工作空間輸入影像
Relational 相關的
Operator 操作者,運算子
Constant 常數
Source 來源,訊號源
Logic and Bit Operations 邏輯和位元運算
Input image type 輸入影像型態
Constant value 常數的數值

2-2
Rotate 旋轉
Background 背景
Fill 裝滿

Gain 增益
Geometric 幾何學的
Interpolation 內插
Free 自由的
Neighbor 鄰居
Center 中心
Conner 角落

2-3

Algorithm 演算法
Unsigned 無號數
Assemble 組合
Display 展覽
Similar 相似的
Conversion 轉變
Variable 變數
Command 指令
Represent 代表
Intensity 強度

2-4

Component 成份,分量
Mother board 主機板
Following 以下的
Bus 匯流排
Require 需要
Instruction 指令
Crack 裂縫,破解
Chroma 色度
Sampling 抽樣
Transmitter 發射器

3-1

Rotate 旋轉
Forward 向前的
Fall 秋天,泉
Number 數字
Electronics 電子學
Circuit 電路
Free running 飛奔執行
Close 接地的
Semiconductor 半導體
Optics 光學

3-2

Resize 重新調整大小
Factor 因素
Label 標籤
Strong 強的
Port 港口
Charge 電荷,費用
Pixel 畫素
Shape 形狀
Rectangular 矩形的
Circular 圓形的

3-3

Crop 取出
Selector 選擇器
Dimension 尺寸
Start 開始
Check box 檢核方塊
Stand 站
Icon 圖示方塊
Busy 忙碌的
Exist 存在
Current directory 現在的目錄

4-1

Hit 命中,撞
Speed 速度
Display 展覽
Resistor 電阻
Structure 結構
Operator 操作者
Architecture 建築
Operation system 作業系統
Memory 記憶
Indicate 指出

4-2

Point out 指出
Demonstration 示範
Example 例子
Remember 記得
Transistor 電晶體
Amplifier 放大器
Block 方塊
Model 模型
Graphic 圖型的
Interrupt 中斷

5-1

Add 增加

Subtract 減去
Review 複習,檢討
Minimum 最小
Maximum 極大
Divide 除,等分
Edge thinning 邊緣窄化
Detection 發現,偵測
Multiple 多樣的,多重的
Integer 整數

5-2

Keyboard 鍵盤
Prompt 提示
Density 密度
Liquid 液體
Capacitor 電容器
Transform 轉換
Voltage 電壓
Direct 直接的
Enlarge 擴大
Extend 延伸

6-1

Conversion 轉變
Simulation 模擬
Toolbox 工具箱
Sink 汲,吸入器,汲極
Filter 過濾器,濾波器
Preference 偏好
Coefficient 係數
Control 控制
Main 主要部份
Sharp 銳利的

6-2

Individual 個別的
Noise 雜訊
Pepper 胡椒粉
In addition 除此之外
Remove 移除
Original 最初的
Noisy 有雜訊的
Salt 鹽
Perfect 完美的
Assemble 組合

6-3

Target 目標
Trace 追踪
Correlation 相互關係,相關
Information 資訊
Identification 確認,辨識
Recognition 承認
Presentation 發表
Introduction 介紹
Realize 了解
Framework 架構

7-1

Distribution 分配
Statistics 統計
Possibility 可能性
Histogram 直方圖
Reshape 改造
Multiplexer 多工器
Regular 正規的,規則的
Screen 螢幕
Adjust 調整
Pixel 像素

8-1

From Multimedia File 從多媒體檔案
Image From File 從檔案中取出影像
Independent 獨立的
Peak 峰值
Data Type Conversion 資料類型轉換
Vibration 震動
Touch 接觸
Panel 面板
Computation 計算
Signal Attributes 訊號屬性

9-1

Double-click 雙點擊
Represent 代表
Transmission 傳輸
Submatrice 子矩陣
Subsystem 子系統
Before 以前
Synthesis 綜合,合成
Transfer 轉移
Bandwidth 頻寬
Architecture 建築,建構

10-1

Amplitude 振幅
Frequency 頻率
Phase 相位,階段
Modulation 調變
Deviation 偏差,偏移
Efficiency 效率
Bandwidth 頻寬
Demodulator 解調器
Scope 示波器
Carrier 載波

10-2

Additive 添加劑,加入的
White 白色
Gaussian 高斯
Noise 噪音
SNR: Noise to Signal Ratio (SNR： 訊號雜訊比)
Channel 通道
Hold 把握,保持
Discrete 不連續的,離散的
Communication 溝通
Cutoff 截止的

10-3

Sideband 旁帶
Index 索引,指數
Execution 實行
Processor 處理器
Media 媒體
Analyzer 分析器,分析者
Spectrum 頻譜
Interference 干涉
Signal processing 信號處理
Simple 簡單的

10-4

Sawtooth wave 鋸齒波
Square wave 方波
Sinusoidal wave 正弦波
Analog 類比的
Digital 數位的
Order 順序
DSB AM: Double-Sideband Amplitude Modulation (DSB：雙旁帶調幅)
DSBSC AM: Double-Sideband Suppressed-Carrier Amplitude Modulation (DSBSC：雙旁帶抑制載波調幅)
Success 成功
Continuous 連續的

讀者回函

讀者回函

GIVE US A PIECE OF YOUR MIND

感謝您購買本公司出版的書，您的意見對我們非常重要！由於您寶貴的建議，我們才得以不斷地推陳出新，繼續出版更實用、精緻的圖書。因此，請填妥下列資料(也可直接貼上名片)，寄回本公司(免貼郵票)，您將不定期收到最新的圖書資料！

購買書號： **書名：**

姓　　名：________________

職　　業：□上班族　□教師　□學生　□工程師　□其它

學　　歷：□研究所　□大學　□專科　□高中職　□其它

年　　齡：□ 10~20　□ 20~30　□ 30~40　□ 40~50　□ 50~

單　　位：________________ 部門科系：________________

職　　稱：________________ 聯絡電話：________________

電子郵件：________________

通訊住址：□□□ ________________

____

您從何處購買此書：

□書局 ________ □電腦店 ________ □展覽 ________ □其他 ________

您覺得本書的品質：

內容方面：□很好　□好　□尚可　□差

排版方面：□很好　□好　□尚可　□差

印刷方面：□很好　□好　□尚可　□差

紙張方面：□很好　□好　□尚可　□差

您最喜歡本書的地方：________________

您最不喜歡本書的地方：________________

假如請您對本書評分，您會給(0~100 分)：________ 分

您最希望我們出版那些電腦書籍：

請將您對本書的意見告訴我們：

您有寫作的點子嗎？□無　□有　專長領域：________________

歡迎您加入博碩文化的行列哦！

✂請沿虛線剪下寄回本公司

Give Us a Piece Of Your Mind

博碩文化網站　http://www.drmaster.com.tw

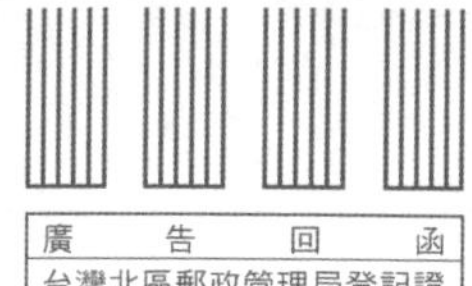

廣告回函
台灣北區郵政管理局登記證
北台字第4647號
印刷品・免貼郵票

221

博碩文化股份有限公司 產品部

台灣新北市汐止區新台五路一段112號10樓A棟

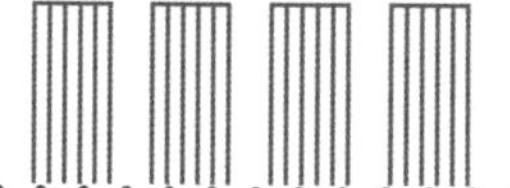

信用卡專用訂購單 CREDIT CARD

※優惠折扣請上博碩網站查詢，或電洽 (02)2696-2869#307
※請填妥此訂單傳真至(02)2696-2867 或直接利用背面回郵直接投遞。謝謝！

一、訂購資料

	書號	書名	數量	單價	小計
1					
2					
3					
4					
5					
6					
7					
8					
9					
10					
			總計 NT$		

總　計：NT$ ________ X 0.8 = 折扣金額 NT$ ________

折扣後金額：NT$ ________ + 掛號費：NT$ ________

＝總支付金額 NT$ ________ ※各項金額若有小數，請四捨五入計算。

「掛號費台北縣 70 元，外縣市（包含台北市） 80 元，外島縣市 100 元」

二、基本資料

收 件 人：________ 生日：____ 年 ____ 月 ____ 日

電　　話：(住家) ________ (公司) ________ 分機

收件地址：□□□ ________

發票資料：□ 個人 (二聯式) □ 公司抬頭 / 統一編號：________

信用卡別：□ MASTER CARD □ VISA CARD □ JCB 卡 □ 聯合信用卡

信用卡號：□□□□□□□□□□□□□□□□

身份證號：□□□□□□□□□□

有效期間：________ 年 ________ 月止

訂購金額：________ 元整 (總支付金額)

訂購日期：____ 年 ____ 月 ____ 日

持卡人簽名：________ (與信用卡簽名同字樣)

黏 貼 處

博碩文化網址

http://www.drmaster.com.tw

請沿虛線剪下寄回本公司

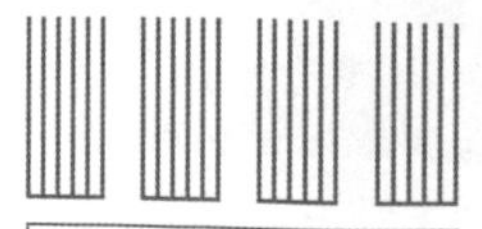

廣　告　回　函
台灣北區郵政管理局登記證
北台字第4647號
印刷品・免貼郵票

221

博碩文化股份有限公司　業務部

台北縣汐止市新台五路一段112號10樓A棟

如何購買博碩書籍

全省書局

請至全省各大書局、連鎖書店、電腦書專賣店直接選購。

（書店地圖可至博碩文化網站查詢，若遇書店架上缺書，可向書店申請代訂）

信用卡及劃撥訂單（優惠折扣8折）

請至博碩文化網站下載相關表格，或直接填寫書中隨附訂購單並於付款後，

將單據傳真至 (02)2696-2867 。

線上訂購

請連線至「博碩文化網站 http://www.drmaster.com.tw」，於網站上查詢

優惠折扣訊息並訂購即可。